W9-AAC-112

Georg Rüppell

BIRD FLIGHT

Georg Rüppell

BIRD FLIGHT

VNR VAN NOSTRAND REINHOLD COMPANY
NEW YORK CINCINNATI ATLANTA DALLAS SAN FRANCISCO
LONDON TORONTO MELBOURNE

My Wife

This book was originally published in Germany under the title Vogelflug

Translation by Marguerite A. Biederman-Thorson
Library of Congress Catalog Card Number 76-58380
ISBN 0-442-27197-2

Printed in the United States.
Published in 1977 by Van Nostrand Reinhold Company
A division of Litton Educational Publishing, Inc.
450 West 33rd Street, New York, NY 10001

Van Nostrand Reinhold Limited
1410 Birchmount Road, Scarborough, Ontario M1P 2E7, Canada

Van Nostrand Reinhold Australia Pty. Limited
17 Queen Street, Mitcham, Victoria 3132, Australia

Van Nostrand Reinhold Company Limited
Molly Millars Lane, Wokingham, Berkshire, England

16 15 14 13 12 11 10 9 8 7 6 5 4 3 2 1

Library of Congress Cataloging in Publication Data

Rüppell, Georg, 1940-
Bird flight.
Translation of Vogelflug.
Bibliography: p.
Includes index.
1. Birds—Flight. I. Title.
QL698.7.R8313 598.2'1'852 76-58380
ISBN 0-442-27197-2

Contents

Preface

A sparrow is chasing a feather, as it falls from the window of a house. Skillfully the bird follows the capricious course, and having caught this bit of building material, whizzes straight up to its nesting place under the eaves.

This is an everyday occurrence—for the sparrow. But if the casual observer knew all that was happening at the moment the feather was captured, he would, I think, be amazed. Every bit of the sparrow's body is in motion. Metabolism, muscles, nerves, and above all the sense organs are operating at top speed—and with beautifully coordinated precision. A few grams of animal matter suffice to direct and execute maneuvers that man-made devices, however well-equipped with the latest circuitry and propulsion mechanisms, cannot imitate.

We humans are ordinarily incapable of discerning all that takes place when a bird flies. The wings move too quickly, and the bird is close enough to be observed for only an instant. Scientists have been studying bird flight for ages, but only recently—with the development of modern photographic apparatus—has it been possible to make the motion of a flying bird so clearly visible that it can be analyzed.

A first step has been to obtain an exact description of all the movements a bird makes in flight. This was achieved by slow-motion films and by photos taken with brief exposure times. Once the sequence of movements was established, the principles of aerodynamics supplied analogies that gave a picture of the forces acting on the bird's wing. This in turn has made it possible to estimate accurately the work done and the energy consumed during flight.

But before all this could be accomplished, extraordinary difficulties had to be overcome. Birds, of course, do not always fly within the range of cameras. Imagine the problems in discovering the technique and energy expenditure of a bird fighting its way through a hurricane over a raging sea!

Occasionally, however, it has been possible to duplicate natural conditions in experiments. For example, in order to find out the effects of the low temperatures and atmospheric pressure (only half the pressure at sea level) encountered by birds flying at an altitude of 6000 meters, sparrows were exposed to similar conditions created artificially. The result was astonishing. The birds' body temperatures fell by only 2°C, and they could fly as well as before. In contrast, white mice under the same conditions became motionless, after their temperatures had fallen by 12°C.

It is really surprising how little is known about bird flight. After all, for millennia people have enjoyed watching and studying birds. Humans have always felt a special relationship with birds. Most birds are active during the day, like man, and apart from a few exceptions they rest at night. Also, like man, birds walk on two legs and are predominantly visual animals, relying more upon the eyes than other sense organs for their perception of the environment. Somehow birds are fundamentally appealing to us; they are not poisonous like scorpions and snakes, nor do they feel cold and slimy like toads, salamanders, or snails. The appearance and song of a bird are pleasing, and its flight, in apparent defiance of gravity, has inspired man to reckless attempts at emulation.

Men have tried again and again to lift themselves into the air with birdlike wings, and they still try. But the structure of the human body simply does not meet the demands of such flight, and no amount of training or exertion can overcome this defect. Birds, on the other hand, are born with the capacity for flight. They need not learn to fly, as the scientist Spalding demonstrated more than 100 years ago. He raised swallows in clay tubes so narrow that the birds could not move their wings. Nevertheless, once they were set free, they flew like their siblings raised under ordinary conditions. From birth birds have the basic ability to fly, but they must practice to develop the finer skills of flight. Young storks can easily be distinguished from their elders by the awkward way the young birds land. Even mature birds must keep fit by constant training. Swifts, for example, are highly specialized for prolonged flying. When they have been prevented from flying for a period, as the result of an accident, their muscles deteriorate and they become quite incapable of flight. Since they must fly in order to catch the flying insects upon which they feed, to be grounded means certain death.

The ability to fly, like all other biological phenomena, came about as an adaptation to particular conditions in a variety of habitats. Since these conditions are different everywhere, today we are confronted with an incredible diversity of flight techniques. Whereas swifts spend most of their lives in the air, other birds fly very rarely and some do not fly at all. The claim that no two bird species fly in exactly the same way is certainly justified.

Why can't a sparrow glide? Why must small birds beat their wings so rapidly? Why can large birds fly faster and longer than small birds? How do migratory birds manage to navigate, with neither map nor compass and without feeding, over thousands of kilometers of ocean to a small island?

One question after another. This book gives some of the answers we have found and offers insight into the most fascinating form of locomotion that has been evolved by living beings on this earth.

A vital element, of course, is the description of the most important aspects of bird flight by means of pictures in which single instants during the flight process are captured on film. They alone allow the fleeting motions of birds to be fixed in time and space, so that we can examine them.

The author will be pleased if his book stimulates the reader to carry out his own observations. Every observation, however small, contributes to a better understanding of the world around us—and such an understanding is essential to its preservation.

Erlangen, Autumn 1974 Georg Rüppell

Chapter One

Everything That Flies

The Effect of Gravity

Everything is subject to the force of gravity. Each time we grasp an object, take a step, or make a movement of any kind, gravity shapes our actions. It must be taken into account when a house is built or a piece of furniture designed. Gravity is so much a part of our lives that we hardly think about it. Not even when a cup falls to the floor are we explicitly aware of what force has smashed it to bits. The plans and materials used in constructing houses, towers, garden fences—almost everything, in fact—are chiefly determined by the effects of gravity. And so it is with the structure of plants and animals. During the evolutionary construction of living organisms, gravity laid down the guidelines.

Organisms that do not move about, such as plants growing on solid ground, are less affected by gravity. The only gravitational problem these organisms face is that of supporting their own weight. Eucalyptus trees and giant sequoias can take their time, many thousands of years, to grow into stable towers built of tons and tons of matter.

Animals, on the other hand, which locomote by pushing themselves away from the ground and landing again, are limited in the size to which they can grow. Elephants cannot gallop nor rhinoceroses jump, because they could not hold up their own weight during such activity—their legs would simply collapse. By contrast, small animals can jump distances greater than the length of their own bodies. The leaps of small antelopes, for example, can reach heights two or three times their body length. If a 6-foot man were to accomplish the same thing, he would have to jump 12 to 18 feet high. The small crustaceans called beach fleas leap 30 to 40 times the height of their bodies, and real fleas as much as 100 times. In equalling this feat, a man would be jumping 200 meters into the air. Why is it impossible for him to do this?

The answer lies in a simple calculation. The weight a leg can support depends upon the cross-sectional area of its bones, and the force it can exert in jumping is also determined by the cross-sectional area of its muscles. Very small animals such as shrews, with little weight for the legs to bear, have quite thin leg bones and muscles. The leg bones of large animals must be much thicker. Now, the weight of an animal is proportional to its volume, and as the linear dimensions of animals increase, volume increases as the cube while the supporting area increases only as the square. As animals grow larger, the point is soon reached at which bones and muscles can no longer support the weight of the body. This is the reason that most of the gigantic dinosaurs lived in the water. And the largest animals alive today, the whales, have evolved in an aquatic environment.

The force of gravity limits growth particularly on land. But it can be used to advantage as well. For example, the shimmering blue leaf beetle, threatened on its exposed alder leaf by the approach of a bird, can save its life only by falling instantly into the grass below. We can easily test whether other insects and spiders make similar use of gravity to find shelter from predators. We need only give a slight shake to the plants on which these animals are active; immediately they drop to the ground. It is to their advantage that the small friction as they pass through the air has only a slight braking effect, so that they fall quickly.

Hatchet fish (Gasteropelecus, *below) and flying fish* (Exocoetus, *bottom of page) have enlarged pectoral fins. The animals on the opposite page glide by means of stretched membranes; these are between the toes of the frog* Rhacophorus *(top) and on the sides of the body in the flying dragon* (Draco, *middle) and flying squirrel* (Pteromys, *bottom).*

"Not-Quite" Flight

The surprised predator is rarely able to catch prey that drops so suddenly out of sight. Capture becomes even more difficult if the hunted animal manages to cover long distances in the air. For example, fresh-water fish must find it useful to make even short leaps over the water surface when they are being chased by pike or perch. It is not at all uncommon to see fish shooting suddenly out of a river or pond. The force of their beating fins catapults them out of the water. Neither these fresh-water fish nor the salmon—which makes high and broad jumps of several meters—possess organs specialized for such feats. But there are specially equipped fish, called halfbeaks, which can jump much further. The halfbeak has an asymmetrical tail fin, with the lower lobe considerably larger than the upper. The fish is propelled from the water by vigorous strokes of this fin. Once under water again, it repeats these tail strokes and goes on hopping for long distances over the surface.

The real flying fishes, on the other hand, can stay in the air much longer by spreading their large pectoral fins. The larger these fins, the longer the fish can sail through the air. Fish of one flying-fish genus have pectoral fins only about a third the length of the body and can make only short jumping flights. But the flying fish *Exocoetus volitans,* which has pectoral fins almost as long as its body, can glide for distances of up to 100 meters. First it gets up momentum under water, then it breaks through the surface and increases the rate of its tail strokes. Because air is less resistant than water, the body of the fish

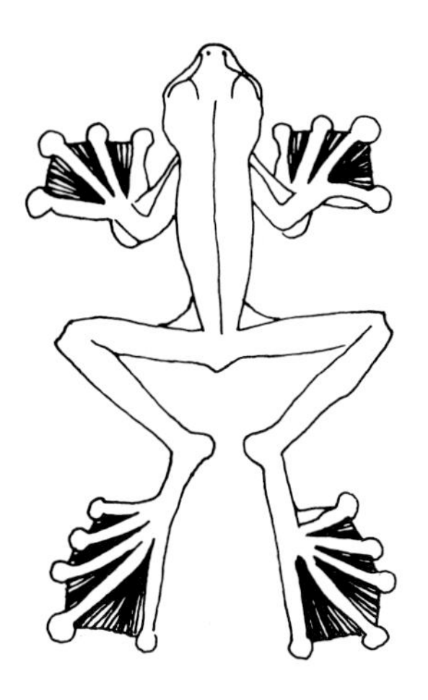

accelerates; it hurtles 10 to 20 meters forward along the water surface, the tail whipping back and forth. Then the fish suddenly spreads its pectoral fins, soars upward, and glides gradually down again. Having touched the water surface, it can repeat the take-off movements and fly again. The longest gliding flight of this sort observed so far, interrupted by brief phases of acceleration along the surface of the water, was 400 meters. Although the pectoral fins of flying fishes can distinctly be seen to vibrate at the take-off, it is doubtful that they actually contribute to forward propulsion.

Other fish have succeeded in propelling themselves with the pectoral fins, using a rowing motion. The South American hatchet fish, only a few centimeters long, respond to a dangerous situation by springing from the water, beating the pectoral fins up and down like wings. Like birds, these fish have developed an especially strong musculature for the purpose, which is lacking in all other fish. These muscles are attached to the large breastbone and account for a quarter of the total body weight. The flight technique of the hatchet fish can carry them only over a distance of 3 to 8 meters.

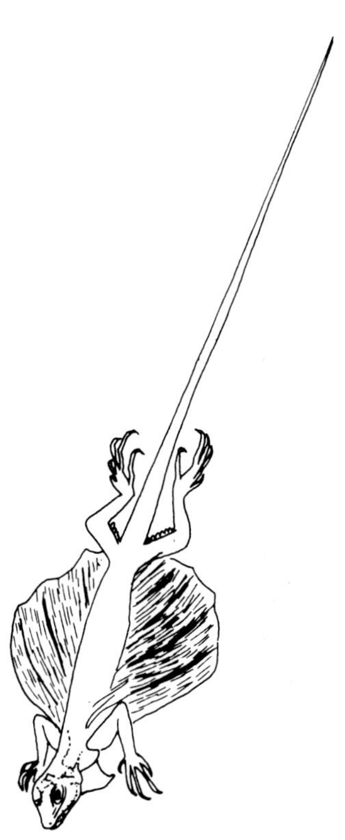

A particularly effective method of escaping from predators by jumping into the air has been developed in certain cephalopod genera. These marine molluscs locomote on the principle of jet propulsion. With powerful ring-shaped muscles the animal expels a stream of water from an opening at the end of the body. The force so generated drives the animal 3 or 4 meters out of the water, like a rocket. At the peak of its trajectory it spreads out the fins along the sides of its body and glides slowly back down. These lateral fins are present in many cephalopod genera and are used in swimming; here they have acquired an additional function as airfoils.

Special gliding organs, in the form of membranes, have been evolved by certain frogs, geckos, squirrels, lemurs, and the lizard called the "flying dragon." These animals leap into the air from trees or rocks, spread out the membranes, and so fly over considerable distances—as much as 70 meters, if the take-off point is high enough and the wind favorable.

Other wingless animals can travel even further, not by using their own body structures but rather by clinging to floats. These are the dwarf spiders, which become dispersed in the autumn by extruding a thread of silk so light that the slightest air movement sweeps it along—and the spider with it. In this way the animals are carried over many hundred meters, even over kilometers, by the wind. After the spiders have landed, the threads can be seen hanging on bushes and trees, sometimes in remarkable numbers.

The dwarf spiders use air travel as a means of locomotion, and the

On the following pages: the brown pelican (Pelecanus occidentalis), *a symbol of aerial superiority.*

other animals that have "almost" mastered flight also take advantage of it to escape from their enemies. But of course the method usually works only against predators that cannot themselves fly. Now we shall consider the many animals in which the technique of flight has been perfected.

The Multitudes of True Fliers

Birds, insects, and bats have succeeded in counteracting gravity by generating forces that support them in the air—they can truly fly. The greatest success of all, at least in terms of survival and diversification, has been achieved by the insects.

Of the million or so animal species on earth, about 750,000 are insects! And this number will probably be increased by many thousands of new species that have not yet been discovered. Furthermore, the population density of insects is extraordinary. To form a picture of this multitude of creatures, consider that in just a few square kilometers of a favorable habitat, there are more insects than humans on the entire earth!

It is certain that the explosive invasion of almost all habitats by insects became possible only after they had evolved the ability to fly; there are no barriers in the air. At first there were no predators in the air either. For 100 million years insects were undisputed masters of the aerial regions—until, about 150 million years ago, the birds evolved.

A June beetle during vertical take-off.

The feathered fliers, however, have developed only about 8600 species. The insects had long since taken the opportunity to become dispersed without interference. What special characteristics can the insects have had that enabled them to spread through the world so much more successfully than the birds?

One of these must certainly be their small size. Many small organisms can live in a space, and with a food supply, sufficient for only a few large animals. Moreover, small animals do not take as long as large ones to grow to maturity. As a result, successive generations appear more rapidly; opportunities for the exchange of genetic material are more frequent, so that genetic changes occur at a higher rate and adaptation to the demands of the environment is facilitated. The small size of the insect body was actually brought about by structural inadequacies. The chitinous exoskeleton cannot exceed a certain size, or it will become too heavy; the respiratory system, which consists of fine branching channels, can function adequately only if these air-filled tubes are not too long.

Small size also involved a disadvantage for the insects; it made them vulnerable to attack by birds. In the effort to escape these predators, insects have developed various forms of camouflage and modifications of flight technique. Watching flying insects, one soon notices that their paths tend to be erratic, leading up and down, back and forth. The apparently uncertain, fluttering course taken by butterflies, caddis flies, mayflies, and other insects makes it difficult for a pursuer to keep the target in view and seize it. Each change of direction forces the hunter to correct his course accordingly, and this can be done only with a certain delay. These sudden, unpredictable changes of direction in themselves offer good protection from predators. Even more effective is the almost instantaneous reversal of course performed by a fly or dragonfly.

Another method of evading predation by birds, adopted by many insects, is the nocturnal habit. By being active only at night they avoid encounters with most birds, which are unable to see in the dark, but they are still exposed to hunting bats. As punctually as though they were changing shifts in a factory, these flying mammals relieve the daytime predators at their work. Bats orient in the dark by emitting short ultrasonic cries and, with their large ears, detecting the echoes reflected by all sorts of objects, including flying insects. In this way they obtain precise information about the position, velocity, and nature of the objects around them.

As skilled as bats are at flying, their technique is not nearly as highly perfected as that of the birds. Membranes are simply not an adequate substitute for feathers. In addition to insects, birds, and a few mammals, we know of but one other group of animals, now extinct, that succeeded in overcoming the earth's force of attraction—the pterosaurians. In Mesozoic times, about 150 million years ago, when reptiles dominated the earth, there were flying forms (*Rhamphorhynchus*, for example, and *Pteranodon*) with a wingspread of as much as 8 meters. Their wings, like those of bats, consisted of tough membranes stretched out between the elongated fingers. *Pteranodon* probably spent its time gliding over the open sea. As the saurians became extinct, these flying forms also vanished from the earth.

Dorsal view of a mouse-eared bat (Myotis). *The head of this flying insect-hunter is not visible; it is being held forward for echolocation.*

The extinct pterosaurian Rhamphorhynchus.

Apart from these four animal groups, only one other organism has attempted to lift itself into the air with its own muscle power—man.

Homo sapiens' Attempts to Fly

Can you swim? A little. Dive? Poorly. Run? Not for very long. Jump? Well... Climb? It depends.

We humans are Jacks-of-all-trades. Our moderate command of the various modes of locomotion is sufficient, however, to bring us to most of the world's habitats: high mountains, deep caves, dense forests, swamps and deserts, the ocean floor, rivers, and lakes. Only the air is beyond our reach. A smattering of skill in this case is not enough. Either one can fly and stays in the air, or one cannot and crashes to the ground.

The inaccessibility of this realm has given rise to dreams, wildly imaginative schemes, more serious but no less dangerous proposals, and eventually the invention of flying machines. But before men succeeded in generating the power needed for flight in machines, the sky was absolutely closed to them. The ability to fly was the sole prerogative, in poetry and legend, of man's gods and other supernatural beings. For centuries the sky has been populated by winged gods and angels, spirits and devils, and the human heroes of classical tales. The dream of human flight is as old as human culture. The Chinese emperor Shun, who lived more than 4000 years ago, was said to have been able to fly and so were figures in Indian mythology and Greek and Roman sagas. The famous Icarus, who flew so close to the sun with his home-made wings that the wax melted and he fell from the sky, was the creation of a viewpoint just as unrealistic as were the ideas about flight that arose during the Age of Enlightenment.

More than one project involved direct use of flying birds. It was claimed that Alexander the Great wanted to have himself raised in a basket by hungry eagles, coaxed to fly upward with pieces of meat. In the 17th century a number of equally fantastic schemes were proposed. For example, in Grimmelshausen's "Simplicissimus" (1684) there is a picture called The Flying Traveller, showing a man hanging from a framework that is carried by birds and driven forward by a sail. Today we are amused by such a suggestion, but the basic idea was correct; that is, a lifting force (the birds) and a means of propulsion (the sail) are required for successful flight.

The notion of flying with one's own wings has, time and again, tempted the human imagination. How great a part sheer guesswork

chau den tollen Wan
dersmann
Und sein rares Flüg
werck an
Denckt was
nicht tan

These ancient Egyptian pictures of birds gliding and flying with beating wings were drawn five to six thousand years ago.

played in the attempts to put these ideas into practice is shown by the tragicomic experiments of the "Shoemaker of Augsburg" and the "Tailor of Ulm." The first of these crashed with his wings of wrought iron just as inevitably as the other, with wings of unsupported cloth. All these visionaries based their efforts on a false assumption—that birds use their wings to "shovel the air downward" and thus stay aloft. No bird could fly by this principle.

It was not until the Montgolfier brothers made their brilliant discovery that hot air, which rises because it is less dense than cooler air, can carry men that the dream of human flight was realized, with the first balloon ascent in 1783. But the voluminous, slow balloons—and the later dirigibles—failed to turn to use the really crucial elements of bird flight: aerodynamic lift, speed (increased by streamlining), and even exploitation of the downward pull of gravity. Balloon flight, based on the lighter-than-air principle, proved to be a dead end in human aeronautics.

Learning from Nature

In the end, it was from the birds that men copied the art of flying. But several thousand years of observation preceded the eventual success. Ancient Egyptian drawings show that birds in flight were studied as long as 5 or 6 thousand years ago. The accurate representation of specific flight maneuvers in these pictures shows how closely and carefully the artists must have watched their subjects. Flying birds appear repeatedly in works of art produced during the following millennia. But it was not until 1500 A.D. that a convention-disdaining genius sought the scientific basis of bird flight. The drawings of Leonardo da Vinci illustrate certain particulars of bird flight that have been rediscovered only in this century. With no theoretical knowledge of the forces that generate lift, he inferred them and used them to explain complicated flight maneuvers. His sketches of aerial turns reveal a talent for uncommonly precise observation. Along with these analyses of movement, Leonardo made detailed studies of bird structure. From such insights he drew plans for the construction of flight apparatus. Designs for helicopter-like machines and airplanes, as well as a statement of the parachute principle, were produced by this most brilliant man of his day.

Drawings on the left: the trick by which Alexander the Great is said to have wanted to fly (top); the Flying Traveler suspended by birds and driven forward by a sail (middle, from Grimmelshausen's "Simplicissimus", 1684); the unsuccessful tailor of Ulm (bottom, 1811). The drawing to the right of these shows the Belgian shoemaker de Groof (1874) crashing to earth after his movable-wing apparatus was raised by a balloon.

About 150 years after Leonardo's time, the Italian mathematician Borelli developed a new theory of bird flight. If a wedge is driven into wood, the wood is split. Conversely, the wedge is pushed out if the pieces of wood on either side are pressed together. When a bird flies, he suggested, the wing is analogous to the wedge, and the air, to the pieces of wood. The air presses on the wing, which is pushed aside and thus pulls forward the body attached to it. This so-called wedge theory happens not to apply to bird and airplane flight, but it is certainly graphic and at least seeks causes and effects. In another respect, however, Borelli made a correct observation. Downward bending of the tail causes a flying bird to pitch downward. Borelli was able to verify this observation with a model in a tank of water.

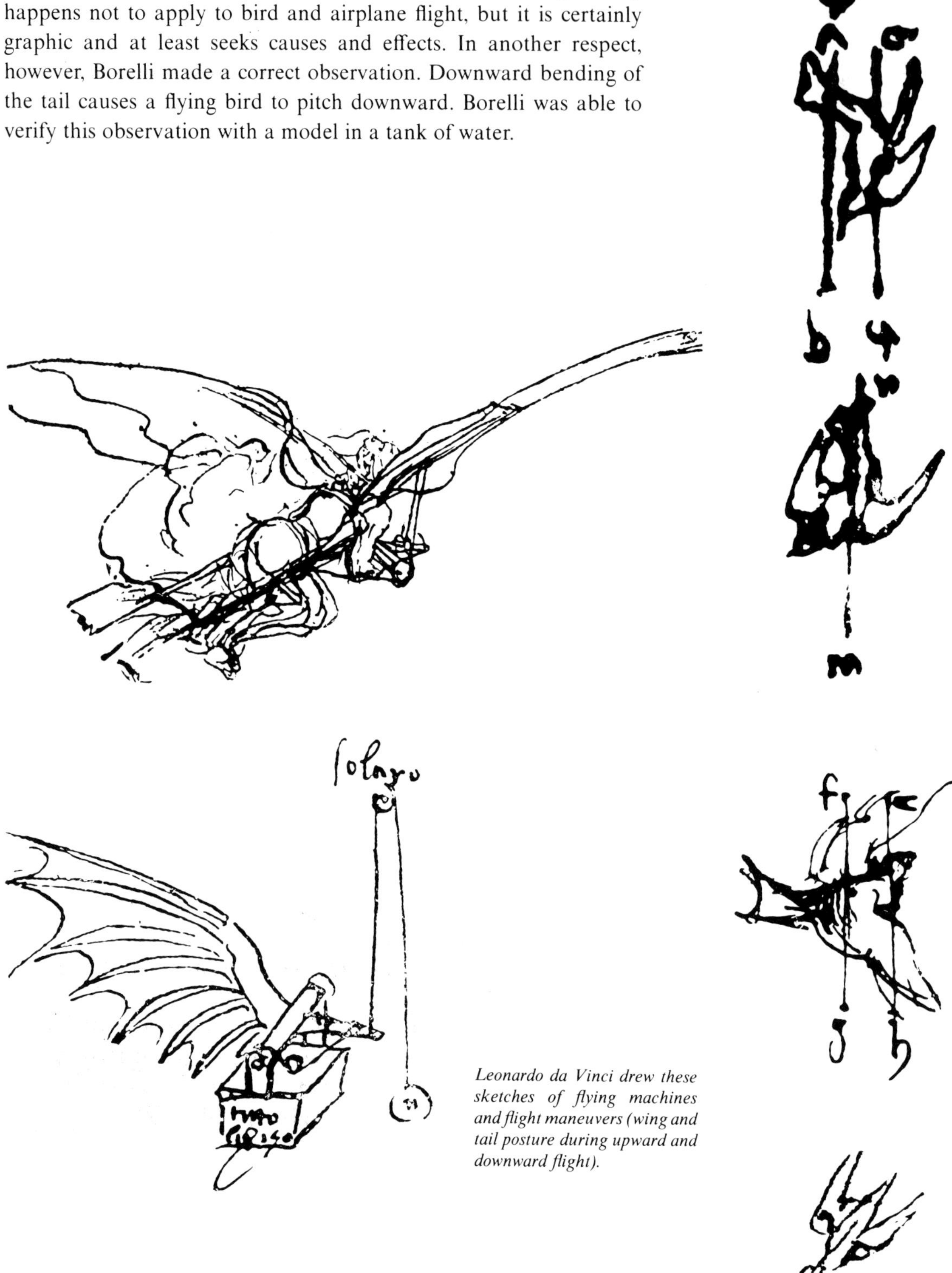

Leonardo da Vinci drew these sketches of flying machines and flight maneuvers (wing and tail posture during upward and downward flight).

In the first half of the 19th century, the English scientist Sir George Cayley got much closer to the secret through his physical experiments and by studying animals. He demonstrated that a bird feather set at an appropriate angle can generate lift when it moves through the air or when air flows around it. Applying his discoveries, he was the first person who managed to remain in the air briefly by means of a heavier-than-air device. His countryman Henson tried in 1842 to build steam engines, intended to drive propellers, into aircraft with of lift-producing surfaces. But all these engines were too heavy. Another Englishman, Graham, was the first to announce the discovery that curved surfaces can generate lift. He was also the inventor of the wind tunnel. Many remarkable "airplanes" were built around this time in Europe, particularly in England and France. None of them, however, could fly.

It remained for the German, Otto Lilienthal to become the first to fly over long distances. In more than 2000 gliding flights from hills, he tested his theory that moving, appropriately curved surfaces could bear the weight of a man, if they were large enough. He developed a special apparatus to test different types of wings; the wing was mounted on a rotatable framework in such a way that he could measure exactly the effects of the airstream. Many surfaces, curved in various ways, were tested and the results presented in diagrams so well conceived that similar ones are still used today. Lilienthal drew upon the results of his experiments to propose an explanation of bird flight. His book, *Der Vogelflug als Grundlage der Fliegekunst* ("Bird Flight as a Basis for the Art of Flying") became a standard reference work, used by all the pioneers of flight.

After the epoch-making discoveries of Otto Lilienthal, airplane builders throughout the world produced a fantastic array of flying machines. The invention of the gasoline engine in the second half of

Otto Lilienthal in one of his gliders.

the 19th century provided them with a suitable source of power. The situation developed into a race between airplane designers in France, England, and America. It was finally won by the Wright brothers, who accomplished the first motorized flight in 1903 and in the following years continued to hold the lengthening distance record.

Men lost interest in flying by the power of their own muscles once efficient motor-driven airplanes were invented. Unfortunately, this redirection of interest occurred before the laws governing flight in the biological sphere had been clarified. The number of researchers and the financial expenditure devoted to working out the mysteries of animal flight seem modest indeed in comparison with the great effort made to develop human flight technology. Bird-watching remained an interesting occupation, but bird flight—having for millennia been

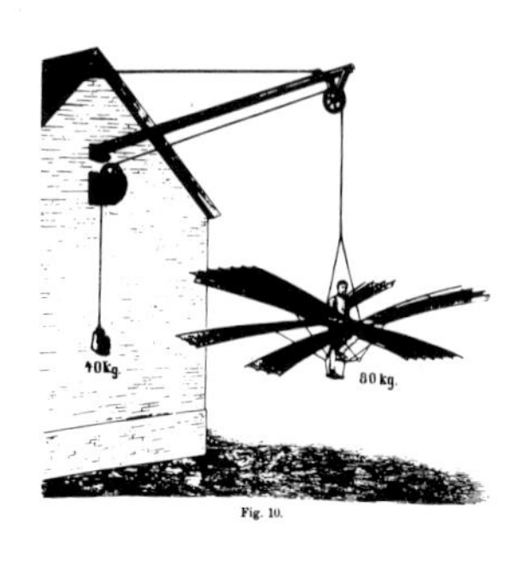

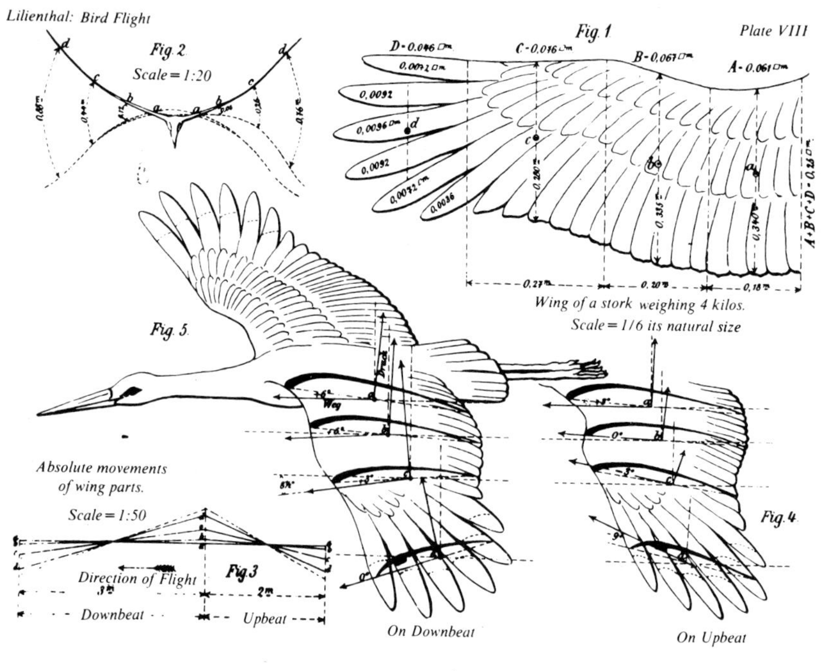

Some of Otto Lilienthal's experiments: beating wings generate lift (above); an analysis of stork flight (left, above); reduction of wing area of a pigeon by tying the feathers together (left, below).

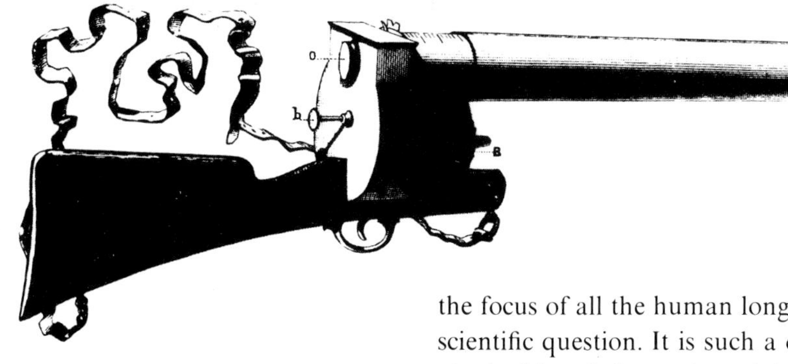

Marey's photographic gun.

the focus of all the human longing to fly—suddenly became a purely scientific question. It is such a difficult area of study that results are obtainable only by costly and time-consuming experimental methods.

For example, the Frenchman Etienne Jules Marey, a contemporary of Otto Lilienthal, had to devise entirely new methods in order to establish the sequence of movements made by flying birds. His results are outstanding, even in the present context. Anyone who has tried to photograph animals knows how hard it is to take pictures of birds in flight. Marey obtained perfectly sharp photographs of flying birds, showing different phases of the motion in rapid succession. To do this he built a "photographic gun" in which a photosensitive plate is moved past a lens in steps and illuminated at brief intervals (up to 11 times per second). This procedure is still used today, though in a refined form, for similar studies of flight.

The crucial realization that the time course of a movement can be analyzed only by its successive spatial representations led Marey to photograph birds from three sides simultaneously. Not until the complexity of the movements was thus revealed did it become apparent how difficult it would be to understand bird flight on the basis of air flow. Another of Marey's great achievements was the invention of methods of recording certain processes in the bird itself during flight, for example, muscular activity. Pressure-measurement devices were connected by tubes to an apparatus that made a record of the frequency and strength of the muscle movements. Marey's research not only produced admirable results in the field of bird flight but provided a model for physiological research in general.

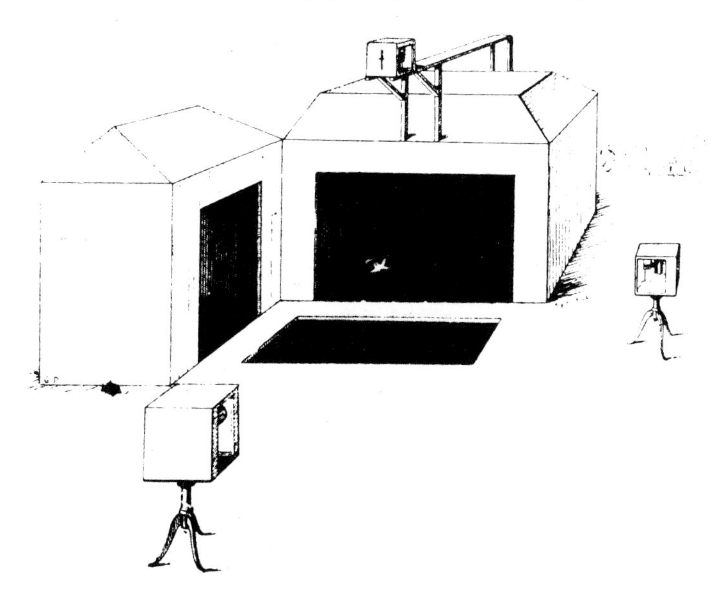

Arrangement of the three cameras Marey used for a three-dimensional analysis of bird flight.

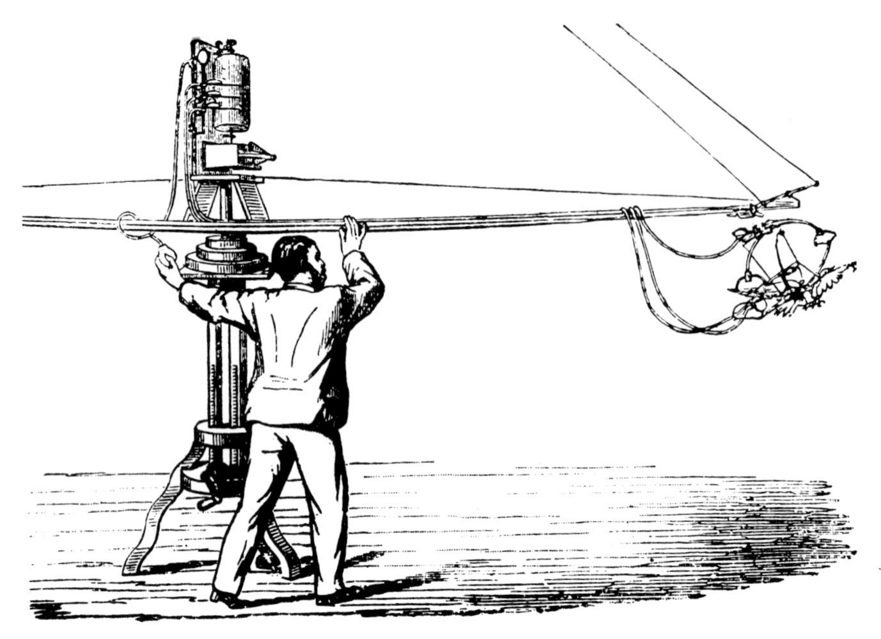

Marey's carousel, to which flying birds were attached by the tubes of the recording apparatus.

After efficient high-speed cameras had been developed, it became possible to learn more about bird flight. In 1940 Stolpe and Zimmer, with special photographic equipment of this sort, presented the first films of small birds in flight.

After 1940, as physicists studying fluid dynamics formulated airfoil theory more precisely, scientists like Erich von Holst, Küchemann, and Oehme applied these formalisms to the problems of air flow around the bird and the resultant generation of forces. If the air-current situation is to be understood thoroughly, it is important to know the precise shape and the exact sequence of movements of the wings and feathers. Nachtigall and his colleagues, as well as Oehme, measured the profiles of bird wings. Tucker and Pennycuick did wind-tunnel experiments on gliding pigeons, falcons, and other birds. Nachtigall and Bilo, by time-consuming analysis of high-speed stereo photographs, managed to work out the many subtle changes a wing undergoes as it beats.

Observation and analysis of freely flying birds also helps to explain the deceptively simple mechanisms of bird flight. The phenomena of natural bird flight were studied 40 years ago by the great student of nature, Konrad Lorenz. Other researchers recorded the sequences of movements and measured the aeronautical performances of flying birds in their own habitats, using high-speed cameras.

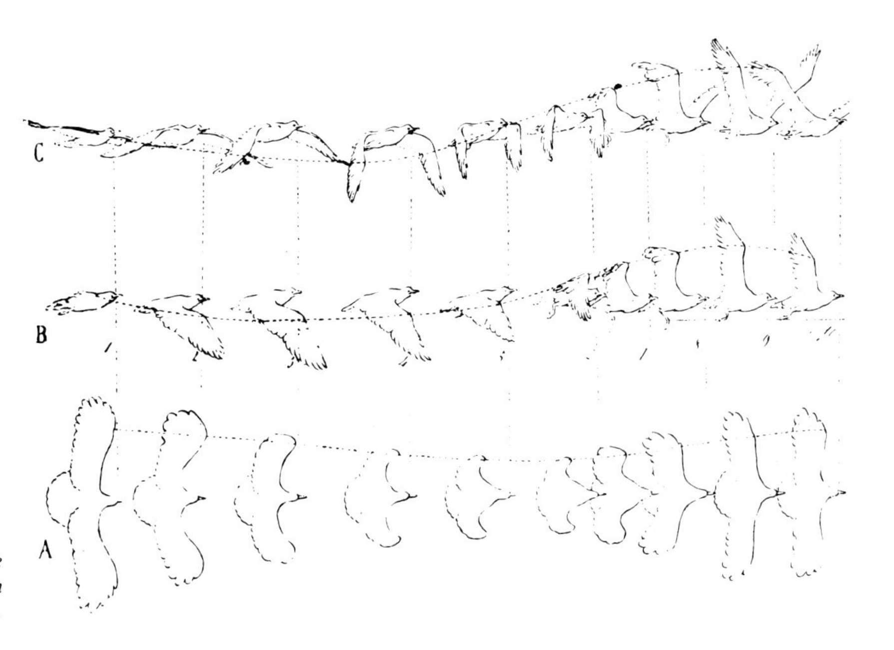

These phases of flight were drawn by Marey from photographs.

The point of all this research, of course, has been to obtain a better understanding of the bird as a flying animal. Birds display unique accomplishments, the details of which are known in only a very few cases. For example, they fly over incredible distances with a minimal expenditure of energy; they turn in a flash; and they fly at altitudes as high as several thousand meters and in the strongest winds. It is clear that they must have an effective metabolic organization, their sense organs must function superbly, and their muscles must be extraordinarily efficient. Whenever one of the as yet unsolved questions about the performance of flying birds is answered, our comprehension of natural phenomena is extended. The observations of behavioral scientists have revealed aspects of bird flight which, as the underlying physiological processes are examined, will help to clarify the web of relationships of which our world is composed.

Chapter Two

The Anatomy of Birds

Uniformity, for a Good Reason

How do we know that the whale is not a fish? Fish have blood that stays at the temperature of their surroundings, and most of them lay eggs from which the young then hatch. Whales, on the other hand, are warm-blooded animals and, like most mammals, bring their young into the world alive.

And how do we know that the cuttlefish is not a fish? Because it has no bones, as is true of all molluscs. Fish, by contrast—as we know from our dinner plates if not from zoology books—have skeletons.

Both of these animals might easily have been taken for fish, though, because they swim in the ocean and resemble fish in shape. In the case of birds, the group to which they belong is immediately obvious, because of one unique characteristic; they cannot be confused with anything else. Everything with feathers is a bird. The presence of two legs, two wings, and a tail, as well as a head with a beak, complete the description. An animal with this combination of features you will identify as a bird, even if you have no zoological training at all.

The modification of the forelimbs into wings has had fundamental effects on the evolution of the body as a whole. Many operations performed by the front legs of the reptilelike ancestors of the birds have had to be taken over by the beak. Because the habits and food of

Left page: In the margin, examples of different body shapes; from top to bottom, vulture, woodpecker, magpie, ostrich, kiwi, hornbill. The photograph shows pink-backed pelicans (Pelecanus rufescens), *with bills modified to serve as fishnets.*

birds differ so greatly, a great variety of beak forms developed; some beaks are shaped like spoons or forceps, and others resemble stilettos or curved daggers.

The front legs were no longer available to support the standing or walking bird, and the overall structure was altered accordingly. The compact body is carried on two legs, often delicate in appearance but nevertheless stable in construction, and well positioned under the body to balance its weight.

But there has been an important limitation on this developing model. If they are to be able to fly, birds cannot become too heavy. The maximum size of a bird is far less than that of, for example, a mammal. The weight of the heaviest bird that can fly, the trumpeter swan, is up to 15 kg; this is only 1/300 the weight of a 4500-kg elephant. To make the point even more graphically, consider the following: if we wished to illustrate the smallest mammal (the pygmy shrew) and the largest terrestrial vertebrate (one of the extinct dinosaurs) on a single page, using the same scale for both, the page would have to be ten times as large as this one—with a smaller scale it would not be possible to see the shrew. A page the size of this one is quite large enough, though, for a similar picture showing the smallest and largest birds—the hummingbird and the condor.

Because of these limitations upon their structure, birds as a group are rather uniform—indeed, almost monotonous—in shape. As though nature were trying to compensate for this, the individual species show great diversity in behavior and magnificently varied coloration.

*The wingspan of the largest and smallest flying land birds, the Andean condor (*Vultur gryphus*(and the hummingbird* (Chaetocercus).

Albatrosses performing their nuptial dance, dueling ruffs, and courting birds of paradise are fascinating to watch, and the brilliant colors of other birds—kingfishers, goldfinches, parakeets, the "flying jewels" (hummingbirds), or hornbills with their bizarre beaks—are just as impressive.

Above: examples of different beak forms. Left, top to bottom: flamingo, seed-eating songbird, saw-billed duck, pigeon, avocet; right, top to bottom: spoonbill, insect-eating songbird, falcon, pelican, swift.

The Bony Framework—Lightness and Strength Combined

Crustaceans and insects have rigid exoskeletons of calcareous or chitinous material, which protect them from the rigors of their environment as well as serving to shape and support the body. Armor of this sort suffers from the disadvantage of being heavy and inflexible—a problem that also vexed the knights of old. The bodies of vertebrates, by contrast, are held upright by an internal framework. The bony skeleton supports the body and, because of the way it is jointed, gives it freedom of motion. In birds, the bones must be both light and tough; light, so that the body is not too heavily loaded for flight, and tough, so that they do not break under the force of landing. This is why bird bones are hollow. Only among flightless species, and diving birds that must submerge themselves, does one find exceptions.

Although birds move about with much greater speed and agility than most other vertebrates, their trunk skeletons are not as flexible as those of, for example, mammals. In mammals the locomotory movements of the fore and hind limbs are coordinated. Therefore , the vertebral column must be able to bend; just think of climbing, jumping, or swimming animals like cats, monkeys, antelopes, and otters. The reduced flexibility of the spine in birds is required for flight. Birds must have a shock-resistant construction in order to withstand the impact of landing, and the skeleton must provide firm attachment surfaces for the powerful flight muscles. In addition, the body must be compact, for otherwise the flight could be unstable. So the bird trunk is a capsule made of rigidly interconnected bones. The most striking aspect of this structure is without doubt the gigantic breastbone with its large keel (or crista), which is absent only in certain shallow-breasted, flightless birds like the rhea and ostrich. To the breastbone are attached the mighty pectoral muscles, which drive the wings. The breastbone is joined to the spine by three to nine pairs of ribs, the number varying among the different species. The spinal column is also sturdily built in the trunk region and the vertebrae, more flexibly linked in other vertebrates, are rigidly interconnected.

Right page: Most mammals have elongated, flexible skeletons (cf. the squirrel, left photograph), but that of the bird is compact and rigid (right photograph and X-ray picture). The drawings show the bones of the pectoral-girdle region, from in front of the bird (below) and turned by 45°. A, upper arm; F, furcula, in this species joined to the breastbone (B) by a cartilaginous element; C, coracoid bone; SB, shoulder blade.

The pectoral girdle, which includes the two shoulder blades, the furcula (the "wishbone" of childhood lore), and the two coracoid bones, closes off the capsule in front. In this region the bony skeleton forming the body capsule is slightly flexible; during breathing, the flight muscles and special respiratory muscles expand and compress the front part of the body. Joined to the rigid body capsule are the neck and head, the tail, and the limbs.

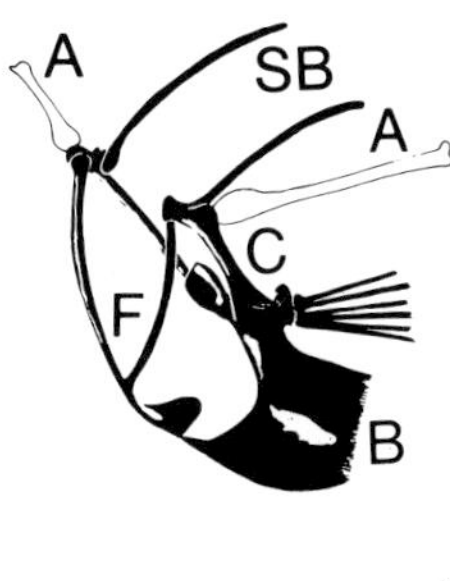

A construction engineer, with the task of building a machine on the basic plan of the bird, would be faced with extraordinary problems. To duplicate the rigid construction of the body capsule would be relatively easy. But to provide it with a movable neck and limbs is a good deal harder. First consider the neck and head: to a body that swings from side to side during walking, and is suddenly accelerated forward or braked, must be attached a self-supporting towerlike structure bearing a weight, the head. Besides simply supporting this weight, the neck must be able to jerk the head suddenly in any direction as well as carry out slow, sweeping, well-coordinated movements, raise loads, and remain motionless while stretched out horizontally for minutes at a time. It must be capable of rotating through almost 360 degrees and of bending far downward. Finally, it must contain passages for the supply of air, water, and food, a

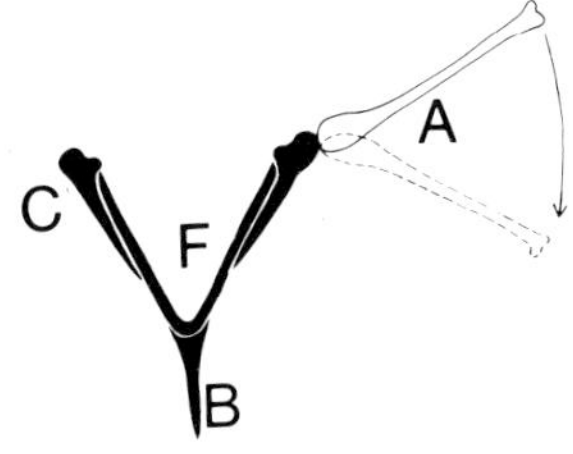

The structure of the bird neck. The S-shaped curve (as in the goliath heron, Ardea goliath, *left) is obscured by the feathers in most birds (cf. the pigeon preparation on the right).*

sound-production apparatus, and an elaborate nervous system mediating a steady flow of information to and from the brain. Look as we may for technological devices that meet all these requirements, we will not find them today.

The bird neck fulfills all these demands. Its many vertebrae and muscles give it great flexibility. Birds are unlike mammals in that the skeleton of the bird neck does not always comprise the same number of vertebrae. A mouse has just as many cervical vertebrae as a giraffe—seven. But the rule for birds is that long necks have many vertebrae (as many as 25) and short necks few, (but at least 11). To increase the range of possible movements, many necks are curved into an S-shape. This curvature is clearly visible in the heron's neck even during flight. Songbirds also have S-shaped necks, but this fact is obscured by the dense contour feathers clothing the neck.

The great flexibility of the bird neck is made possible by the relatively shallow curvature of the vertebral joint surfaces, particularly in the first vertebra, the atlas, which bears the head. A system of long bands of muscle, running the length of the neck, and of other muscles linking each vertebra with the next, operates with fine coordination. Three or four lightning-quick contractions of the neck muscles and a heron has caught a fish. Herons wait until their prey has come within reach of the neck. Then any attempt by the victim to hurl itself out of the danger zone comes too late; the heron's attack is too sudden.

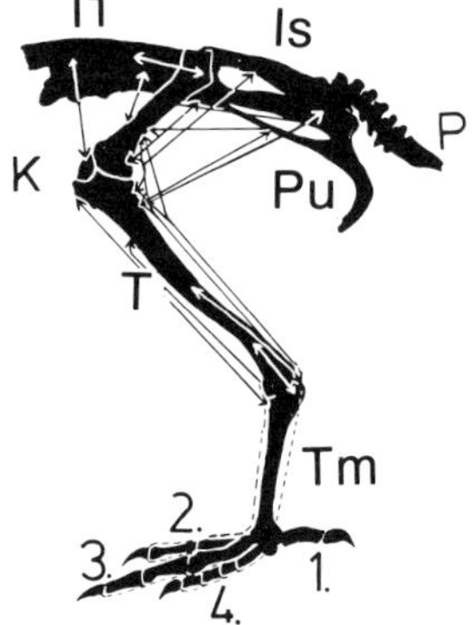

The structure of the bird leg. The photograph shows walking ostriches, with a diagram of the leg to the left. Il, ilium; Is, ischium; P, pygostyle; Pu, pubis; K, knee; T, tibia; Tm, tarsometatarsus; 1, 2, 3, 4, toes. The arrows indicate the direction in which the muscles pull; the dashed lines, that of the tendons.

To hunt successfully, herons must identify their prey at the earliest possible moment. Therefore they, like almost all birds, have extremely good vision. The large eye sockets are a conspicuous feature of the bird skull. But most of the volume of the skull is occupied by the brain. All the messages coming from the eyes and the other sense organs must, of course, be processed or stored there. Moreover, birds have complicated forms of behavior, all of which must be controlled by the brain. To protect the delicate organs of the head, a unique skull structure has developed, with extremely light but strong plates and bridges—fused, in mature birds, with no trace of dividing lines. The outer surfaces of the bony plates are relatively smooth but on the inside are hollow spaces and spongy structures. The toughness of this light skull construction is well demonstrated by the woodpecker: it literally beats its head against the wall!

The skeleton of the tail consists of ten to thirteen vertebrae. The anterior five to seven of these are freely movable, while the posterior five or six are fused to form a triangular piece, the pygostyle. The pygostyle bears the tail feathers and the muscles that move them. During flight birds use their tails primarily for braking and to control the orientation of the body. In addition, the tail acts as a visual signal in threat and courtship behavior. In the flightless birds, which move only on foot, the tail has become almost nonfunctional. But these birds depend chiefly on the wings in courting and antagonistic behavior and also to keep their balance when they walk. They have no real pygostyle at all.

A bird consisting only of a body capsule with the most important organs of metabolism, the head with its sense organs, brain, and beak, and the tail would certainly be able to survive—but only in an automatic feeding box. It may be that chicken batteries will one day produce such birds. In nature, birds must be able to move about if they are to survive. The role of the two legs and feet in bird locomotion varies widely. In the case of pheasants, which walk or run more than they fly, it is essential to have strong and agile legs. But the legs of many birds that are skilled fliers, such as swifts, swallows, and hummingbirds, are almost never used except when the bird perches.

The legs of many birds find uses other than walking and perching. Chickens and other fowl scratch the ground with their feet and fight by slashing with their sharp spurs. Birds of prey have daggerlike claws capable of killing prey; the birds hold on to the prey with these claws just as tits do and can use them as weapons as well. Treecreepers, woodpeckers, and nuthatches drive their sharp claws into tree-bark-like crampons. Swimming birds use their legs as oars. Birds' legs also serve as catapults on take-off and shock-absorbers on landing.

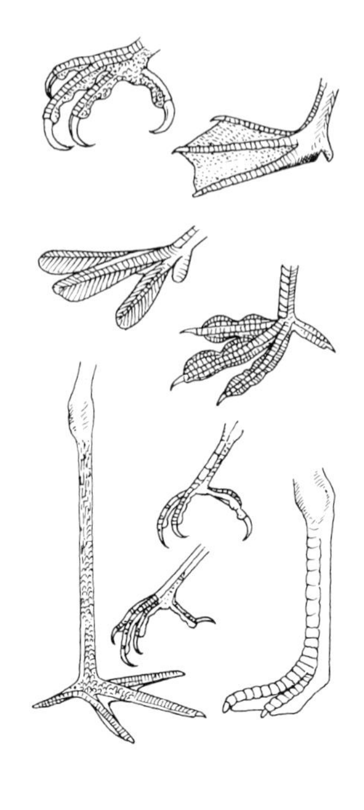

Above: Examples of foot modifications. Left, top to bottom: bird of prey, grebe, stork. Right, top to bottom: goose, coot, lesser spotted woodpecker, songbird, ostrich.

The photographs show feet performing different functions: on the right, hooded vultures (Necrosyrtus monachus) *jumping and fighting, and below, a songbird about to land.*

In spite of this wide range of functions, and despite the differences in external appearance, all bird legs are similar in basic structure. Between the femur (thigh bone) and the tibia (shin bone) is the knee joint. This is reinforced by two ligaments, one on each side of the joint; the structure of the joint capsule itself and the leg muscles, surrounding it on all sides, provide further stability. The thigh is not ordinarily visible, being held close to the body and merging with its contours under the covering of feathers. The uppermost part of the leg that can be seen is the shin—the tibia and small fibula, surrounded by muscles and covered with feathers. The joint at the lower end of the tibia is unique to birds—the intertarsal or hock joint. No four-footed vertebrate has an analogous structure. Since birds are descended from reptiles, we can assume that bird legs have evolved from reptile legs. The hock joint is a new structure; its proximal part is formed by fusion of the tibia with the tarsal bones nearest the body.

The remaining tarsal bones have fused with the metatarsals to form the bone (the "tarsometatarsus") on the other side of the joint, in the "naked" part of the bird leg. Joined to this bone are the toes, usually four in number. These have hinge-type joints so that, like our finger joints, they can bend in one plane only. Examination of the schematic drawing on p. 33 will show that the "new" hock joint of the bird has taken over the function of our knee joint. In addition, the birds have

retained a "real" knee joint between femur and tibia. Why should this be? Perhaps in order to increase the ability of the leg to absorb the impact of landing. When force is applied simultaneously to several joints connected by elastic muscles and ligaments, the load is distributed among them. Furthermore, shifting of the center of gravity when the body tilts forward or back can be compensated for by slight adjustments at the joints. The degree of bending is regulated by muscular contractions; the tensions of the muscles hold the joints in position. The muscles and ligaments together function rather like the cables operating a crane.

There is also an automatic mechanism enabling birds to sleep on their perches. When my son falls asleep, the hand clutching his toy car opens, and the toy clatters to the floor. But when a bird has gripped a branch with its feet, its grasp is not released even during sleep. The tendons running to the toes are arranged in such a way that they are under tension when the hock joint is bent. The joint always bends as the bird settles down, so that the feet automatically curl around the unstable branch. The highly specialized structure of the tendons is correlated with this function; the inner surfaces of the tendons bear little hooks which engage in a cross-ridged tendon sheath.

I once had a good demonstration of how well this gripping mechanism functions, when observing a heron trying to fly away during a violent storm. In order to present the smallest possible surface area to the wind, it tried to take off from a crouching position. But in that posture the hock joint was bent, and for several seconds the heron tried in vain to break its toes' automatic hold on the branch. Not until the bird raised itself up on extended legs did it succeed. Then the wind seized it; it lost its balance in the air and spun downward for a few meters. The presence of stormy winds turned the otherwise well-designed automatic gripping system into a trap for this bird.

Structure and Function of the Wings

Only airplane wings are made of metal. Birds' wings consist of bone, muscle, tendons, nerves, connective tissue, and many feathers. The bony framework has undergone a number of modifications during its evolution from reptilian foreleg to bird wing. For one thing, the bones of the hand have regressed. Birds have only a few bones in the outermost part of the wing skeleton. This is called the hand part of the wing or "manus," since it is homologous to the foot of a reptile or the human hand; the inner part of the wing skeleton, consisting of the humerus, radius, and ulna, is called the proximal wing or arm. The

The gripping mechanism of the foot spoils the take-off of a grey heron (Ardea cinerea; *inset). The large picture was taken a moment later; the tail is outspread and tilted back, to bring the bird out of its fall.*

humerus is often quite short. At one end is a large ball joint that fits into a socket in the pectoral girdle. This especially well-developed joint is necessary to provide the wing with its great mobility. There are special ridges along the humerus for the attachment of the massive flight muscles. The lower arm, as in four-footed vertebrates, consists of radius and ulna. Its length varies over a wide range in different bird species.

What movements does this bony framework permit? In contrast to four-footed animals, birds cannot bend the manus up and down; it is capable only of lateral rotation in the plane of the hand. When the elbow joint is extended, the radius and ulna slide past each other. This shift in relative position presses the two lower-arm bones against the wrist joint in such a way that the manus is extended. Because of the special orientation of the joint surfaces, the wing turns so that the leading edge is angled upward in the region of the arm bones and downward in the region of the hand bones.

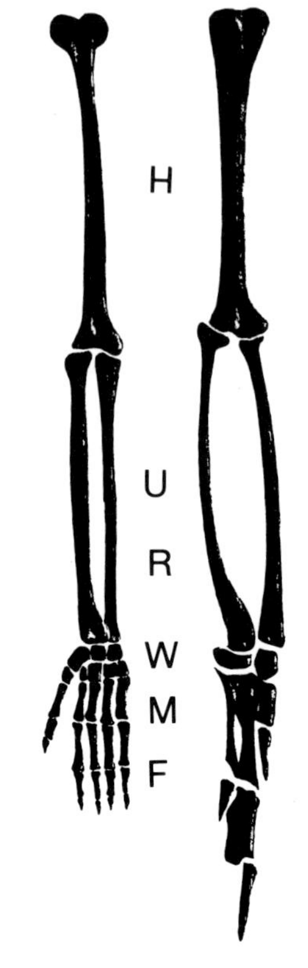

Comparison of the arm and hand bones of man and bird. H, humerus; U, ulna; R, radius; W, wrist; M, metacarpals; F, fingers.

Now, how is the wing as a whole moved? When the large pectoral muscle contracts, the wing is pulled down. The antagonists of this muscle, the small pectoral muscle and the deltoid muscle, raise the wing. The large pectoral muscle does most of the work. Accordingly, it is the largest muscle the bird has; it fills almost all the space between the keep of the breastbone and the base of the wing. Some will know it as the most rewarding mouthful in a roast Cornish game hen. The small pectoral muscle lies beneath the large muscle. When it contracts, its tendon, running through a hole in the bone and fastened to the upper side of the arm bone, pulls the wing up. The effective direction of force here is obtained just as it is in a crane, where the equivalent of the contracting muscle and its tendon is the motor pulling a cable down. Because the cable runs over a pulley, it acts to pull the load upward. The analogous lifting action of the small pectoral muscle is reinforced by contraction of the deltoid muscle, which is so located that it exerts an upward pull directly. The main function of the deltoid muscle, though, is different; it must rotate the wing back out of the leading-edge-down position into which the large pectoral muscle has turned it. During the upward stroke of the wing, the deltoid muscle turns the leading edge upward. This muscle also functions in other ways. Its anterior part bends the elbow joint and also stretches the skin between shoulder and wrist. These three muscles—the large and small pectorals and the deltoid—provide most of the driving force for moving the wing. The muscles responsible for the great flexibility in posture of the wing are smaller ones in the region of the shoulder and upper arm (which also assist in the fine control of wing movement) as well as the muscles in the wing itself,

Right page: The wing in the photograph has been partially plucked to expose the anterior membrane, which is supported by a tendon to form a taut leading edge during flight, and the marginal ligament. The third, posterior wing membrane (cut away) stretches between the elbow (E) and the body. Below is a drawing of the muscles of a pigeon; the most conspicuous is the large pectoral muscle.

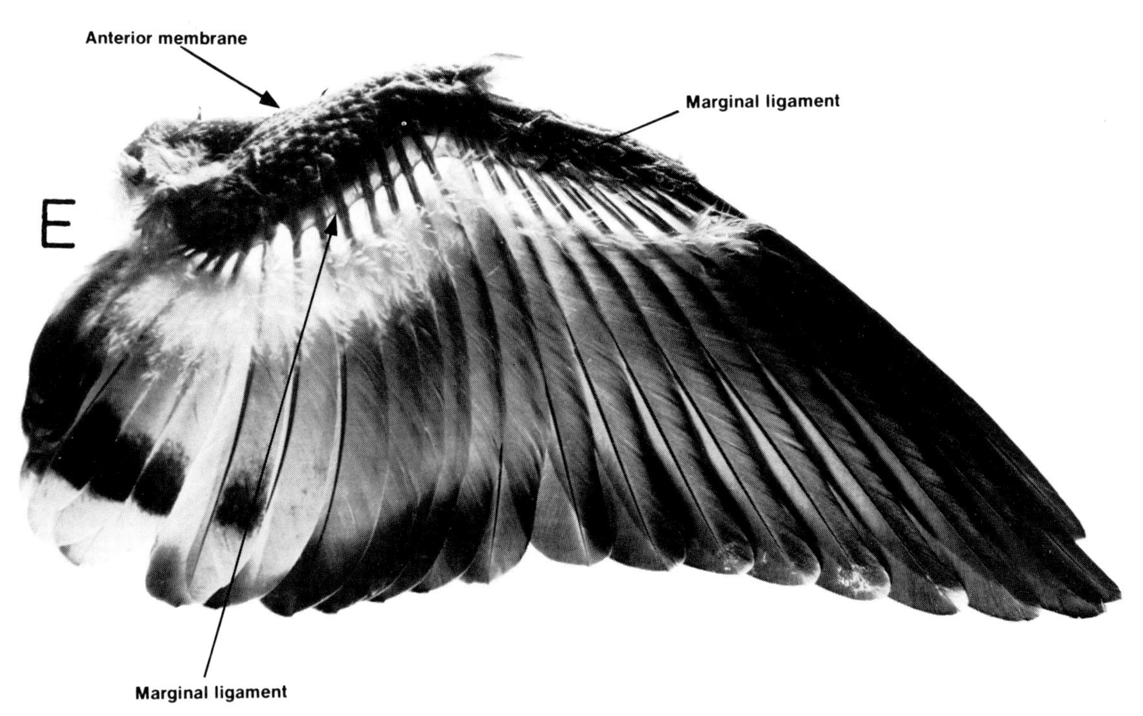

about 20 in number. The importance of the flight muscles is underlined by the fact that in good fliers they make up half to three-quarters of the body weight.

Bones and muscles are not the only components of a wing. To provide an unbroken bearing surface, other structures are necessary—the skin and feathers. The skin in places forms membranes joining the different parts of the wing; these fill in the spaces between the bones and delimit the central part of the wing at the leading and trailing edges. Moreover, it is in the skin that the feathers are inserted. There are three major skin sections on the wing: the anterior and posterior membranes and the large marginal ligament between elbow and metacarpal bones. The bones, the muscles, the membranes, and the connective tissue interspersed between all these structures must be supplied with nutrients. This is the job of the blood vessels, very fine branches of which run throughout the wing.

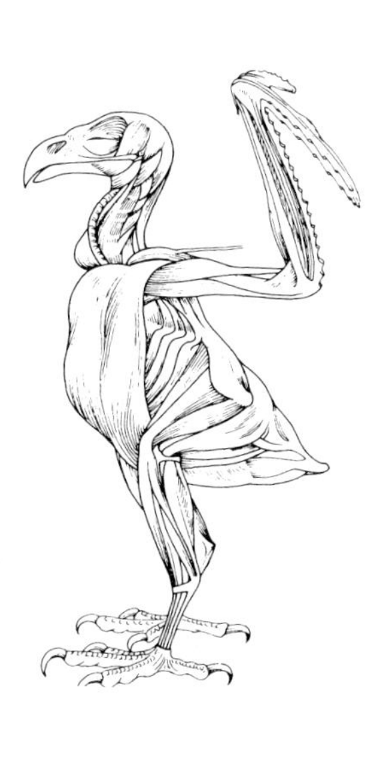

Proper function of this complicated appendage requires coordinating elements, the nerves. These carry the commands of the central nervous system (CNS), which consists of the brain and spinal cord, to the muscles. Other nerves transmit sensory information—coded, like the CNS commands, in the form of electrochemical signals—in the opposite direction, back to the CNS; "extension completed," "high air resistance," "flexion completed," and so on without interruption, as long as the bird is in the air.

The Source of Power for Flight

Have you ever wondered where the power comes from that is needed to carry a bucket, to lift something, or even to walk? A silly question–from the muscles, of course! But how is power generated there? The force-exerting element–the motor–is the muscle, but it operates quite differently from an internal-combustion motor. In the latter, explosions in the cylinders generate a force which eventually, by way of the pistons, rods, and transmission, moves the wheels. By contrast, in muscles nothing is pushed apart; rather, the muscles contract. What brings about this contraction? Let us consider the structure of a muscle. In general, it rather resembles a telephone cable: inside a large tube are many smaller tubes, which in turn contain still smaller tubes. The largest such tube encloses a bundle of muscle fibers. The tubular "fibers" are individual cells, each having a nucleus and a membrane. In turn each fiber contains hundreds of even smaller structures, the myofibrils. Thse are but a few thousandths of a millimeter in diameter but are still not the ultimate contractile structures! The electron microscope reveals that these fibrils are in turn made up of thousands of longitudinal overlapping "filaments," each of which is on the order of only a few millionths of a millimeter in diameter.

Here we are at the "macromolecular" level and can examine the actual events of contraction. In ordinary skeletal muscle (the kind used in most locomotion), these filaments are of two types, depending upon the basic protein they contain–myosin filaments and actin filaments. In each segment (or "sarcomere") of a fibril, the myosin and actin filaments overlap. Contraction actually consists of increasing the degree of this overlap.

But how is this change of overlap accomplished? In the electron microscope, each of the myosin filaments looks rather like a very long thin boat, with hundreds of little oars sticking out from its sides. These oars (called "cross bridges") are now known to be extensions of the myosin molecules that make up the myosin filament. The precise mechanism is not yet clear, but most research now indicates that these oars can make temporary connection with the actin molecules in the several actin filaments surrounding each myosin filament in the closely packed array. During each brief connection, a "rowing" action occurs, causing the filaments to slide past one another, which increases the overlap–the muscle contracts.

Two further basic things are needed, a supply of energy and a means of control. The control–the switching on and off of the contraction–is quite complex; on cue from the central nervous system, signals from motor nerves cause calcium ions to be redistributed in the muscle cells, and these in turn switch on the ability of the cross bridges to "connect and row." An energy source is also needed; the movement can occur only in the presence of a certain energy-rich compound, adenosine triphosphate (ATP). ATP must be synthesized from compounds lower in energy. In the process, these compounds are used up in the muscle and must be replaced; a steady fresh supply is brought by the blood. But this transport takes time. That is why there is a limit to the rate at which muscles can do work.

The synthesis of ATP requires nutrients which are first broken down by oxidation. Usually fats and carbohydrates such as sugar are consumed. About one-third of the chemical energy set free in these reactions is in the form of heat, while the remaining two-thrids provides mechanical energy for the contractile process. To avoid overheating, the muscle must be cooled. Furthermore, the waste products formed by the chemical conversions must be eliminated. Cooling and waste disposal are the responsibility of the circulatory system. For this reason, circulation is particularly brisk in animals with highly active muscles. This is reflected, for example, in the weight of the heart; in most mammals the heart makes up about 0.5% of the total body weight, whereas in birds and bats it amounts to 1%. For this circulating blood to be properly effective, of course, it must carry a sufficient supply of oxygen.

The respiratory system of the bird is designed to meet especially great demands. The paired lungs are connected to five pairs of air sacs, which branch throughout the body. Air sacs even invade the bones. They act both to decrease the weight per volume of the animal and as a bellows, by means of which the air is pumped in and out through the tubular channels of the lung. When air passes through the lung only once, as it does in humans, less oxygen can be withdrawn from it than with the two-way flow of the bird lung. But even though the oxygen supply of birds is better than that of mammals, in many cases it is not sufficient. If the muscles are overworked for a long time, there can be an energy crisis. This is particularly a problem for long-distance fliers. To prevent such an emergency, birds with muscles that must make prolonged effort have a built-in oxygen reserve. These contain myoglobin, the red muscle pigment with properties similar to those of the red blood pigment, hemoglobin. Both pigments are capable of combining with oxygen and releasing it when needed fbr the breakdown of molecules to obtain chemical energy. Myoglobin is found in some muscles of all birds and of all vertebrates that perform movements over prolonged periods. Muscles used only for brief periods lack the red pigment and appear white. Chickens usually move about on foot and fly less often; therefore, the leg meat is red and the breast, the flight musculature, is white. It is possible, then, to tell from the color of the musculature whether the muscles of an animal are designed for prolonged work or only brief activity. Many birds must be capable of both; for example, pigeons, terns, redshanks, and crows can take off quickly and also fly long distances. Their flight musculature includes a white component for the sprint and a red component for the long haul.

The duration of flight also determines which nutrient—fat or carbohydrate—is predominantly consumed in the muscles. Carbohydrate is water-soluble and in the dissolved form can be transported more rapidly to the site where it is needed; for the sprinter, therefore, it is the most suitable. Fats, on the other hand, which need not be dissolved in water (a fact associated with some saving in weight), have more than twice as much energy per unit weight ("caloric value") and therefore are the fuel most suitable for long-distance fliers. The breakdown of fats always requires oxygen. Oxygen is not absolutely necessary, though, for the breakdown of carbohydrates; the latter can also be broken down anaerobically. When this happens, the muscle builds up a so-called oxygen debt, which can be paid later by increasing the supply of oxygen.

In most circumstances, the muscles of a bird are capable of generating power sufficient for flight. But a bird constantly pursued—for example, because it has escaped from its cage in a room—must eventually become exhausted and stop for a rest. Then it gasps for breath with an open bill. This indicates that it is probably in an oxygen-deprived condition and heat production in its muscles has progressed to such an extent that the blood is no longer cooling them adequately. When the temperature in the muscle becomes too high, the chemical reactions are no longer properly controlled. Built-in temperature sensors notify the brain that the overloaded thermoregulatory system has broken down; as a result, the brain switches off activity and switches on accessory ventilators—the bird breathes heavily with its bill open. As it does so, liquid evaporates. This evaporation cools the blood as it flows through the tissues.

The Architecture of a Feather

The most important components of a bird's wing are the feathers. It is also possible to fly with membranes, as the flying saurians did and as the bats still do. But it works better with feathers, and birds are amply equipped with these. There are over 1000 feathers on a small songbird, more than 6000 on a gull, 12,000 on a duck, and 25,000 on a swan. If you should ever try counting feathers for yourself, be sure to pick a day when the air is still. Feathers are wonderfully light objects. Despite their lightness they are sturdy, flexible, and easy to care for; they provide a cushion, thermal insulation, a water-repellent cover, and—most importantly—they are replaceable.

The basic construction of feathers is the same in all birds. The shaft, which projects from the skin, is often hollow or filled only with spongy keratin. (Keratin is the horny substance of which the feather is made.) Many branches along either side of the shaft together form a flat surface called the vane. Each of these "barbs" in turn has on each side a row of smaller branches, the "barbules." The barbs are all the same, except perhaps for some variation in length, but the barbules differ according to the side of the barb from which they branch. Those pointing toward the tip of the feather bear hooklike projections, while those pointing toward the feather base are curved and hence may be called "bow" barbules. The hook barbules on one barb engage with the bow barbules of the adjacent barb. Thus, the two barbs are held together, and if the connection should be broken they can easily be hooked together again. (Try smoothing a ruffled feather with your fingers and you will see the effect.) This arrangement is indispensable if the feather is to present an uninterrupted surface. The hook barbules also bear upward-pointing processes. These little hooks roughen the surface of the feather and help to keep the wing surface unbroken by increasing the friction between the overlapping feathers. This is of course particularly important for all those feathers which together constitute the surface that bears the weight of the bird.

Feathers are classified via their different functions. There are usually ten flight feathers in the hand, called the primaries. Often these are sharply pointed and asymmetric, with a very narrow outer vane. In many heavy birds which beat their wings up and down very rapidly, the primaries are strongly arched, with the tips curving downward so that they can be considerably distorted before they are flattened. The adjacent flight feathers, in the ulna region, are called secondaries; they have inner and outer vanes of almost equal size. The secondaries vary greatly in number, depending on the length of the wing. The albatross has 37, whereas most songbirds have only nine or ten secondary feathers. Covert feathers shape the wing profile and generally insulate and protect the wing. The flight feathers of the tail ("retrices") are important in changes of flight direction and in braking maneuvers. "Contour feathers" is the name given to those of the trunk, for they determine the outer, aerodynamically desirable shape of the bird. We can guess the function of the down feathers from their

The fine structure of a feather: The branches at either side of the shaft bear hook and bow barbules which catch in one another like little zippers. The top and center photographs show a covert feather of a goose; the bottom inset is a scanning electronmicrograph of a pigeon flight feather.

use in sleeping bags and quilts. Finally, there are feathers serving chiefly for decoration and display—a property at times put to use by man as well as by the feathers' original owners.

All these kinds of feathers are built in basically the same way. But certain elements may be slightly modified, and their numbers may be increased or decreased. For example, in contour feathers the barbs do not cling together so strongly, and the feathers themselves adhere to one another even less readily. The hook barbules bear few hooks; instead, they are broadened in places so as to provide a shield against external influences like wind and rain.

Down feathers have barbs, but no specialized hook and bow barbules. The side branches of the barbs are short and threadlike.

The incredibly complicated structure of the bird feather is illustrated by the following figures: In a flight feather of a pigeon, about 1000 barbs branch from either side of the shaft. Each of these branches in turn bears about 550 barbules. The total number of barbules in a single feather, then, is enormous—nearly a million.

When the wings are furled the individual flight feathers lie one over the other like shingles. The many air spaces left between them make the whole structure very light and insulate it against heat loss. When the muscles extend the wing, the feathers slide past one another to maintain a thin surface; the wing fans out. Because of the constant friction to which the feathers are exposed, during this folding and opening and by contact with other objects and the air itself, the feathers gradually wear out. But this is no problem for a bird—a critical property of the bird feather is that it can be replaced.

How the Feathers Are Replaced

There are no naked adult birds—unless, of course, they have been plucked and are on their way to the kitchen. But many birds occasionally look as though someone has been trying to pluck them; this happens when they are changing their feathers (molting). The molt of most birds usually recurs regularly, like the breeding season, at intervals of about 12 months. Following an internal, preprogrammed schedule, the basis of which has not yet been discovered, these birds begin to cast off the old feathers and grow new ones. Since the loss of feathers makes it more difficult for them to fly, and since the regeneration of feathers costs energy, birds molt during the least severe seasons, and usually outside the breeding period. In regions where the climate has an annual rhythm, as in Central Europe, molting usually follows breeding. In other regions conditions favorable to breeding can occur quite unpredictably; in such places it is not

Above: the flight feathers on the right wing of this bearded vulture (Gypaetus barbatus) *are bent backward–a good demonstration of their bending strength. Middle four photos: Flattened barbules in the contour feather of a hawk (upper left); goose down (upper right); a pigeon wing (lower left) and its components (lower right): flight feathers, covert feathers, alula (the group of feathers attached to the first digit), and down.*

Below: the short-eared owl (Asio flammeus) *brakes itself by fanning out its tail feathers.*

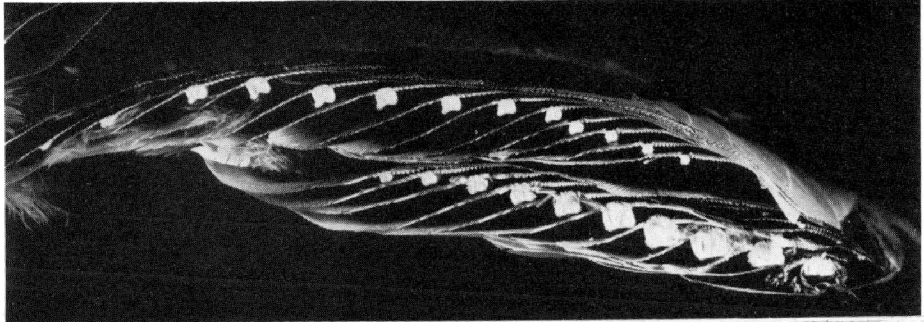

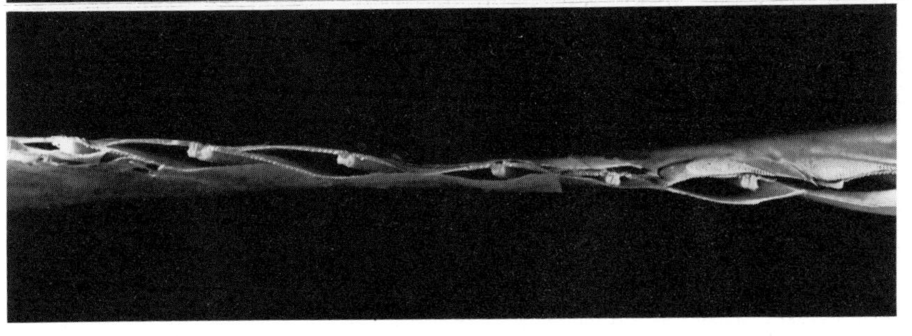

so easy to decide when the time is suitable for molting. Some species of Australian birds, which begin to breed only after the onset of the rainy season, have been observed to interrupt the molt when a rainy period begins unexpectedly and complete the business of breeding first.

What actually happens when molting time arrives? Molting is controlled by the thyroid gland, the activity of which increases well prior to the onset of the molt. It secretes the hormone thyroxin, the messenger substance that announces the beginning of the molt to the feathered areas of the body and wings. Most species capable of flight manage to retain the ability at this time, by not discarding too many feathers at once. The simultaneous shedding of many feathers is

Feathers grow (top left: young white stork, Ciconia ciconia*), wear out (top right: old stork), fall out during molting (middle left: the cormorant* Phalacocorax carbo*), and are replaced (middle right: the marabou,* Leptoptilus crumeniferus*; note sixth feather on right wing). Long-eared owls shed the entire tail at once (bottom left: tail feathers shed, bottom right: feathers replaced).*

feasible only for species that need not depend entirely on flying to escape from their enemies—water birds like auks, loons, grebes, or geese as well as cranes and birds on secluded nests. These birds cast off all the primary feathers at once and for a time cannot fly; when danger threatens they find hiding places or simply dive out of reach.

It is interesting that flightless species, such as the Galapagos cormorant, differ from related species that fly in that they do not renew their feathers at particular seasons. Since these birds cannot fly anyway, it is presumably irrelevant when the feathers are shed. In all species that remain capable of flight during molting the primaries are replaced one after another, in a specific sequence. When the new feathers begin to grow, the bird needs a large supply of material in a short time. The process must be especially rapid in species that lose all the flight feathers simultaneously. In the crane, for example, the new feathers grow as much as 9 millimeters per day. Rapid growth of the new feathers of course, minimizes the time during which these synchronously molting birds are unable to fly. The small duck species can fly again after only 3 weeks, the larger after 4 weeks. Geese can use their wings again 5 weeks after losing their feathers, while swans take 6 or 7 weeks. Species that wear out their wings very rapidly molt twice a year. Among these are long-distance migrants like the willow wren and certain small songbirds; for example, the feathers of the hedge sparrow are severely abraded as it slips through the dense tangle of twigs.

Convex Upward—A Property of All Wings

Whereas the shape of the wing can vary considerably in birds of different species, all wings have one thing in common: the upper surface forms a convex arch. This curvature is visible at a glance, but it can be better examined by cutting through the wing of a dead bird. This reveals the cross-section of the wing—its "profile." Wings may be more or less strongly arched, thicker or thinner. Many more studies are required if we are to understand the detailed connection between form and function—between the nature of the profile and its relationship to the movement of the wing.

The profile of a bird wing is different at each point along its length, from the base of the wing to its tip. Both thickness and curvature of the wing are greatest at the base and decrease steadily toward the wingtip, where separation of the individual primary feathers frequently produces slots or gaps (in which case the wing is said to be "emarginated"). As a wing moves through the air, its profiles change

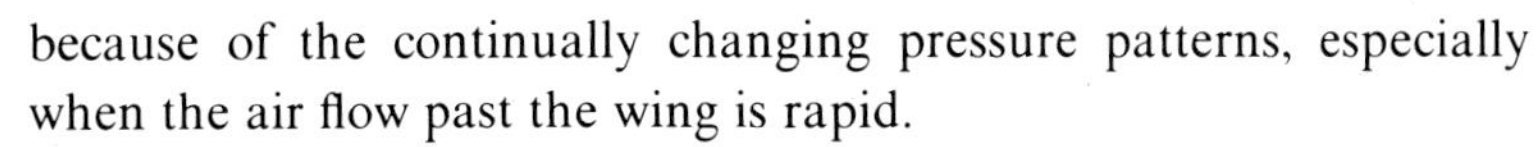

because of the continually changing pressure patterns, especially when the air flow past the wing is rapid.

An expert in the physics of air flow would no doubt complain loudly if he had to calculate the detailed dynamics of bird flight. Where and when should the measurements be made? There are no standardized components or constant quantities in a flying bird. But research demands reliable measurements, numbers with which computation can be done and hypotheses tested. If the dimensions of the bird itself—its outer contours and its weight—are so variable, how can one possibly expect to obtain exact data on the airflow patterns during all the different aspects of flight, which are a thousandfold more variable? In studying bird flight, we must be satisfied with qualitative conclusions. Quantitative, numerically documented knowledge is obtainable for only a few special aspects of the problem. We shall try to analyze the general cause-and-effect relationships in a flying bird. But for the time being we must leave unanswered the question of the precise magnitudes of these causes and effects.

Despite the variety in wing shape all the birds have one thing in common—the curvature, clearly evident in the right wing of the falcon. The photographs show the black-tailed godwit (Limosa limose, *top left), Arctic tern* (Sterna paradisea, *top center), a redstart* (Phoenicurus phoenicurus, *top right), the African white-backed vulture* (Pseudogyps africanus, *middle), and a falcon* (Falco tinnunculus, *bottom). The upper four sketches of wing cross-sections show the variation of curvature in different birds (from top to bottom: sections through the arm part of the wing of the mallard, swift, rook, and redstart, scaled down to the same size). The lower sketches, all from the wing of a blackbird, show the progressive change in curvature and thickness from the base (top profile) to the tip of the wing.*

Chapter Three

What Holds the Bird Up

A Short Course in Aerodynamics

The effects of the earth's gravity, centrifugal force, and inertia are familiar to all of us and often have unpleasant associations. One careless step, and we fall downstairs. Take a curve too fast, and the car skids off the road. When a bus driver suddenly puts on the brakes, the consequences can also be serious, since at the moment of braking the passengers continue to move forward.

The same forces act upon an airplane and upon a flying bird. The exact way that this happens, however, is beyond our ability to observe directly. To explain the origin of lift on a wing, let us try a few little experiments that illustrate the forces the air exerts. If you have held your hand out the window of a moving car, you have already experienced several of these effects. First, let's clarify one common source of confusion about the lifting force on a wing. If you hold your hand in a flat plane (not arched or cupped) like a metal "vane," and tilt it with the leading edge upward, the tilted vane forces air downward and as a reaction your hand is given some lift upward; the effect is like that on a rudder of a boat or plane or the "diving planes" of a submarine. At the same time, you will find that there's a great deal of backward "drag," tending to push your arm back. This "vane" effect is *not* the basic principle by which the airplane wing and bird wing obtain lift!

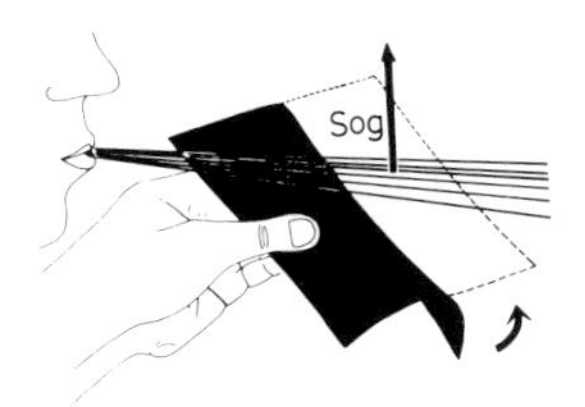

The experiment of blowing over a piece of paper demonstrates the lowering of pressure in a region of rapid air flow.

Put your hand out the window again, and this time hold it slightly cupped so that the upper surface is convex from front to back; at a high enough speed you will actually feel the hand being drawn more strongly upward than back. Here is the force that supports bird and airplane. Turn the hand to a slightly steeper angle, leading edge up, and the lifting force suddenly disappears–the hand is simply pushed backward.

How is this lifting force produced? The flow of air about a moving, convex-upward wing is shown diagrammatically in the illustration on p. 00. One can obtain such a picture by putting a wing in a wind tunnel and tracing the air particles with streams of smoke. The flow is complicated but note that the lines of flowing air particles tend to be crowded together at the upper convex surface. At the lower surface this crowding does not occur. The complex flow thus results in a region of constriction in the flow pattern above the wing. Now, whenever such a constriction is present in a steady-state flow pattern, the velocity in that region necessarily increases.

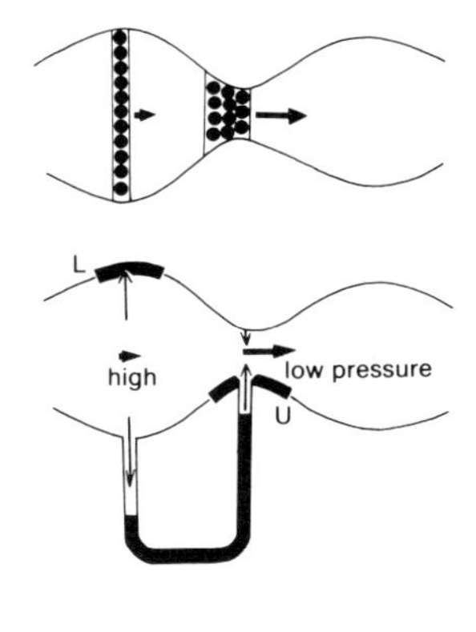

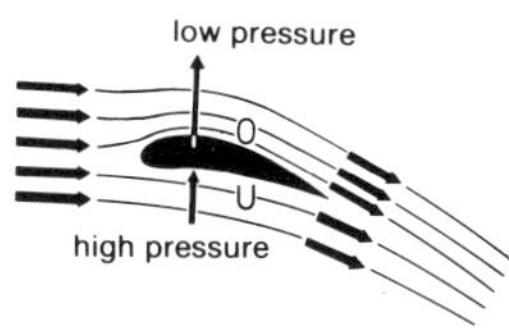

Airstream velocity (proportional to length of the thick arrows) and pressure (direction indicated by thin arrows) in a Venturi tube. The black dots represent air particles. Pressure differences are revealed by the differing heights of the liquid columns in the arms of the glass U-tube. The situation in the parts of the Venturi tube marked U and L corresponds to that on the upper and lower surfaces, respectively, of a wing (lower drawing: wing cross-section with lines of air flow and directions of effective net pressure).

The basic idea is illustrated if you try to blow out a candle with your mouth wide open. It's hard to do. But if you put your lips together to constrict the flow, the velocity of the air—for a given volume displacement—goes way up and the candle is easily blown out. The aerodynamic situation at the wing is much more complicated, but the net result is that the convex-upward wing causes a maintained local region of increased airstream velocity near the upper surface.

Here is another experiment; it shows quite directly that lifting force is exerted when air flows at higher speed above a surface than below it. Hold a piece of paper as in the illustration, and blow over the top of it. The loose end of the paper rises.

But why does this increased velocity at the upper surface cause lift? To understand this effect we must discuss two properties of moving air—static pressure and dynamic pressure. We are all familiar with static pressure, the force exerted uniformly in all directions, like that of the atmosphere or of the air in a blown-up beach ball. "Dynamic pressure" is equal to the kinetic energy (the energy due to motion, proportional to velocity squared) per unit volume of a moving fluid. This is the pressure that one would measure if he were to bring a moving airstream to a standstill—the pressure of the wind against your face is dynamic pressure. The Pitot tube, used to measure speed of an airplane, brings a bit of the airstream to a stop (relative to the airplane), and the dynamic pressure developed in it is used to determine airspeed. Dynamic pressure is also aptly called "stagnation pressure."

Now, there is a fundamental relationship between static and dynamic pressure in the various parts of a flowing airstream, which was clarified by the physicist Bernoulli and is known as Bernoulli's Law. This states, in our present context, that the sum of the static and dynamic pressures must add up to a constant; increase one and the other will necessarily be reduced. There is nothing mystical about this; it's a direct consequence of the laws of conservation of energy (though we won't go into the derivation here).

We can demonstrate Bernoulli's Law graphically by blowing air through a "Venturi tube," as shown in the illustration. The alternating constrictions and expansions in the tube cause local changes in velocity of the flow. In the narrow passages the fluid must flow faster than in the swellings, for in both regions the same amount of air must pass in a given time. From Bernoulli's Law we expect that the static pressure in the constricted parts will be less than in the swellings, because in the former the velocity (and hence the dynamic pressure) is greater. This is easy to show; in the illustration a U-shaped glass tube containing liquid is connected to these two portions of the tube. The fluid rises in the arm connected to the constriction, showing that the static pressure there is less.

Here, then, is the explanation of why the increased velocity above the wing leads to lift. The greater airstream velocity above the wing means a greater dynamic pressure (proportional to velocity squared) there. The static pressure above the wing is thus reduced, for as we have seen the sum of the two must be constant. Since the static pressure above the wing is then less than the static pressure below, the wing is subjected to a net upward force. At sufficient speeds and with appropriate wing shapes and sizes, this force can more than compensate for the weight of a bird or an airplane.

One must be careful not to say that there is an "upward pull" on the upper surface of a wing. The air of course presses against all surfaces of the wing; it is simply that the static pressure is less above than below, which leads to a net upward force on it.

Depending upon shape, it may also be arranged that static pressure beneath the wing can exceed atmospheric pressure—the reverse of the effect above the wing—causing even greater lift. The total upward force produced is greater, the larger the wing surface; as we shall now see, it depends upon a number of other factors as well.

Right page, upper photo: A common puffin (Fratercula arctica) *braking its flight just in front of its nest on a rocky cliff. Lower photo: The almost ideally streamlined body of the glaucous gull* (Larus hyperboreus) *generates very little drag, and the curvature of the wings ensures the necessary lift.*

The Complexity of "Lift," and Other Forces Affecting Flight

So far we have discussed only the lift acting on an idealized horizontal wing. The direction of this net force on the wing due to static pressure differential is at right angles to the direction of the airstream (or "relative wind") flowing over the wing. For an airplane in horizontal flight the relative wind is of course horizontal, and this "wing lift" is directed vertically, counteracting gravity. In the flapping, capricious flight of a bird, however, the situation is obviously far more complex. Even during steady flapping, the wings are angled differently at each instant during the cyclic wing strokes. There are also local air currents, not to mention the fluctuating angles introduced by the acrobatics of climbing and diving. It is clear, therefore, that in discussing bird flight we shall frequently need to refer to the average lift generated by the wings during a stroke or a maneuver. And, as we shall see, though lift ordinarily provides the upward force which overcomes gravity, in some cases the direction of aerodynamic "lift" on the wing may be far from the vertical. Indeed, the net "lift" generated by bird wings must have some forward component in steady flight, or there would be no thrust to propel the bird and overcome the other factors—viscous resistance, stagnation at the leading edges, and other properties of the

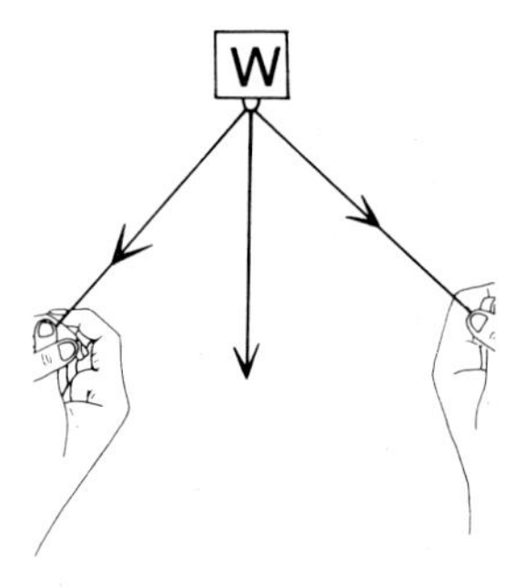

The interaction of forces: if a weight is pulled in two directions, it moves along a line between the two (the diagonal of the parallelogram in a vector diagram).

airstream—which tend to resist such motion (i.e., they contribute to "drag").

In considering the effects of all these forces on a bird, and the net forces which result, vector diagrams are helpful. When two or more forces act in different directions, their overall effect can be described as a single resultant of them all. For example, as shown in the illustration, if two hands are pulling on a weight from different angles, the weight moves along a line between the two directions of pull. The weight will fail to move only if the two hands are pulling equally strongly from exactly opposite directions. If one hand pulls harder than the other, the weight will move toward that hand. Such an interaction of several forces can be diagrammed by means of vectors. A force vector is an arrow; its length indicates the amount of force, and it points in the direction of the force. If two forces are involved, the resultant vector is found by completing the vector diagram as follows. The two vectors form adjacent sides of a parallelogram, and the resultant force is given by the diagonal of this parallelogram; its length represents the magnitude of the resultant force, and its direction corresponds to that of the resultant force.

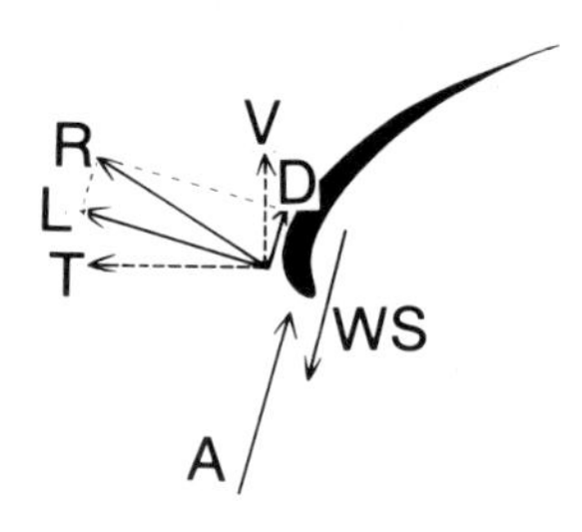

The distribution of forces over a wing profile: at right angles to the airstream (A), lift (L) is produced. A vector diagram representing this force and drag (D) gives the resultant (R), which is composed of a vertical component and a forward-directed component (thrust, T). WS = direction of wingstroke.

Such a diagram can be constructed, for example, to find the resultant upward force due to all of the forces acting on a bird—drag, thrust, lift, and gravity. If the resultant is positive, the bird climbs. For maintained forward flight at constant speed and altitude, there must be an upward force equal to the force of gravity, and thrust must equal drag.

A critical element in this balance of forces is drag. To minimize the energy expenditure necessary to overcome drag, birds have evolved streamlined shapes. But streamlining alone is not enough—the shapes of wings also affect the total effective drag in subtle ways.

Pointed Wings, Low Drag

The total drag acting on a flying bird can be described in terms of several physical effects. First, there is friction between the entire surface of the body and the air streaming past it; this produces surface-friction drag. A second factor is the size of the surface presented to the relative wind, the leading edges of the wings and the front end of the body. The component associated with this factor is called pressure or form drag. At some points on a flying object these contribute to the "profile drag," which is usually defined as independent of the amount of lift generated. One component of the drag at the wing, the induced drag, does depend on lift; it arises as follows. When there are pressure differences between different regions in the

Right page: A characteristic of all birds in flight is illustrated here by the herring gull (Larus argentatus)*; the surfaces generating lift are large (above), and those generating drag are small (below).*

surrounding air, such as between the lower and upper surfaces of the wing, the air tends to flow in such a way as to reduce the difference. That is, it flows from the high-pressure region beneath the wing toward the low-pressure region above it. This compensatory airstream is restricted to the wingtip, since the main airstream associated with flight is too strong at the leading and trailing edges to allow appreciable air flow in other directions. The effect of this compensatory airstream is to push the main airstream toward the base of the wing on the upper surface, and pull it toward the wingtip on the underside. When the upper and lower parts of the airstream are rejoined at the trailing edge, their directions of flow are thus slightly different, and eddies are produced. The eddies produced directly by the compensatory flow around the wingtip are even more pronounced.

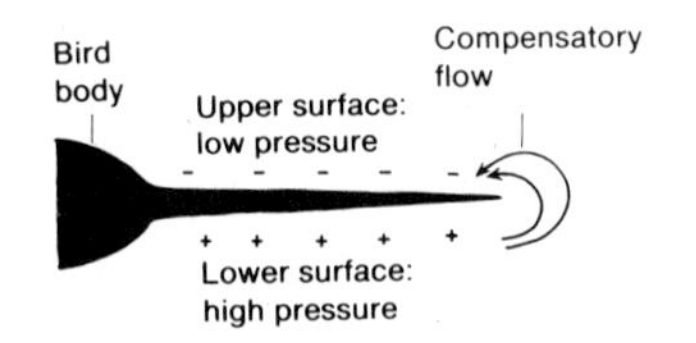

Induced drag is brought about by compensatory airflow around the tip of a wing (arrows on right). The sketch represents the front view of half a model bird.

The energy wasted in maintaining these eddy currents is, of course lost and cannot contribute to the kinetic energy of the forward flight—that is, there is an effective braking of the bird. In order to keep this induced drag as small as possible, bird wings are pointed at the tip, a shape which is taken to an extreme in the wings of albatrosses and swifts. The advantage of a pointed wing, then, is that at its tip the pressure differences are slight (there is little lift, since here the air flows around the wing over only a small area) and the tip area, where the compensatory current can flow, is relatively small.

The aspect ratio, the ratio of wingspan to wing width, is an especially important consideration in designing a wing to minimize induced drag. (For convenience, we shall use the term aspect ratio in discussing bird wings, but in this case base-to-tip length will be meant rather than overall wingspan.) The longer and narrower a wing is, the better, since lift is exerted over a large area compared to the narrow tip, where induced drag is generated. In albatrosses, the aspect ratio is particularly favorable, ranging from 6:1 to 8:1; in sparrows, on the other hand, it is only 3:1. It is possible that vultures and buzzards, the wings of which are very blunt, diminish the induced drag by splaying out the primary feathers so that the wingtip is emarginated. The effect of this arrangement is to make the very outer edge considerably smaller, since then it consists effectively only of the tips of the individual feathers.

Some small aircraft also have special devices to reduce compensatory currents at the wingtip, flat plates mounted on the wingtip at right angles to the surface of the wing. But in the design of most airplanes the induced drag is considered more acceptable than the new problems in controlling the plane that these plates bring. Birds, which must conserve energy as much as possible, go to great lengths to reduce drag—unless, of course, they want to slow down.

Above: to keep the induced drag small, bird wings either end in a point like those of the albatross (Diomedea irrorata) or have a slotted margin as in the rosy pelican (Pelecanus onocrotalus).

Below: with the leading edge of the wing tilted far up, the airstream separates from the wing; the resultant force is almost entirely composed of drag.

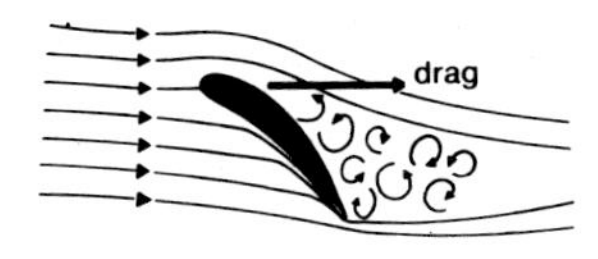

In Braking, Drag Is Desirable

When forward motion is to be slowed, drag must be greatly increased, in order that kinetic energy of motion is dissipated. Birds increase drag by tilting their wings back. The same effect is achieved in airplanes by lowering the flaps. Both methods produce an increase in the angle between the underside of the wing and the incident airstream. As this so-called angle of attack is increased, lift at first stays as large as before despite the reduced airspeed, since at the leading edge, where the air flows over the uptilted arch of the wing, the effective curvature is increased and the velocity of the airstream increased accordingly. In compensation, the volume the airstream can occupy at the trailing edge is greater than before; so it flows more slowly, and the static pressure and hence the pressure drag are increased.

If the angle of attack continues to increase, there is eventually a disruption of the desired airstream, beginning at the trailing edge of the wing and moving progressively forward. As a result, drag, already large because of the greater surface presented to the relative wind, increases still further. As the wing tilts more steeply, more and more of the wing generates drag, and the part generating lift becomes smaller. If the bird is to avoid a crash, it must take care not to make the angle of attack too large.

The Significance of the Boundary Layer

When air flows over a wing, the molecules of air directly at the wing surface are swept along with the wing. These molecules in turn tend to drag along the molecules in layers of air immediately above them, and so on. As a result, there is a very thin layer, called the boundary layer, in which the properties of air flow are very different from those in the main airstream that produces the lift. The thickness and nature of the boundary layer vary with the velocity of the flowing medium, the shape and size of the surface involved, and the viscosity of the medium. With air it may be only a millimeter or so thick.

Here we are mainly concerned with the decisive effects the boundary layer can have on disruption of the optimal airstream. If the flow in the boundary layer is orderly, the particles stream along parallel paths, and there is a smooth diminution of velocity from the main airstream to the wing surface. Such flow is called laminar. At higher velocities, or if the flow is perturbed by the shape or extent of the surface passed over, the flow in the boundary layer can become turbulent. Turbulent flow is disorderly, with the direction of flow changing rapidly at each point; in particular, the particle velocities near the wing surface can be higher than in laminar flow, so that the air within a turbulent boundary layer has more energy and momentum.

How do these properties affect the airflow near a wing? It turns out that, especially at high angles of attack, the viscous drag within a laminar boundary layer can slow the air within it, as it flows over the wing, to such an extent that turbulence and vortices can spread forward from the trailing edge of the wing. The main airstream producing lift can then actually "separate" from the wing surface, greatly impairing its performance. Surprisingly, however, it has been found that if turbulence is artificially induced in the boundary layer early in its passage over the wing, then the separation is much retarded. The greater momentum of the turbulent boundary-layer air carries it farther back along the wing surface, so that larger areas of the wing can generate lift even at high angles of attack.

Now, how can a bird ensure the formation of such a turbulent boundary layer? When an airstream is divided by the wing, the boundary layer is initially laminar. But it can be converted into a turbulent layer if the leading edge is particularly sharp or if rough structures are present (so-called turbulence generators) such as the

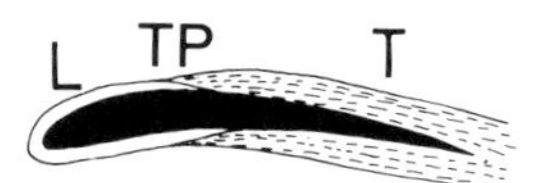

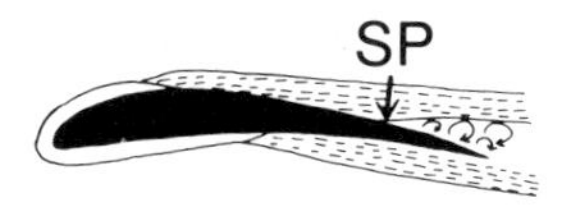

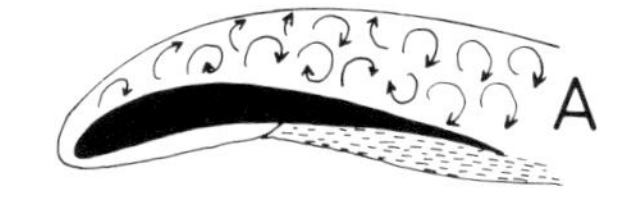

Top sketch: the laminar boundary layer (L) is converted to a turbulent layer (T) at the transition point (TP). Middle: the boundary layer can separate from the wing at the separation point (SP). Bottom: with the boundary layer gone, the airstream (A) is disrupted.

barbs of feathers. The profiles of most bird wings, near the end of the wing, are so sharp-edged that the boundary layer is probably turbulent right up to the leading edge of the wing.

Wings that are round and thick, like those of the fulmar or murre near their bases, are more likely to have a laminar boundary layer. To overcome this disadvantage, such birds must fly at high speed or beat the wings rapidly (which also enhances lift generation). This strategy is effective because at high airstream velocities the disruptive separation of the airstream from the wing is less likely to occur; the boundary-layer air carries more momentum. Moreover, higher velocities favor the formation of a turbulent boundary layer.

In discussions of laminar and turbulent flow, a very useful index of the overall situation is given by the so-called Reynolds number (Re). This is a dimensionless function, calculated as

$$\mathrm{Re} = \frac{v \cdot l}{\mu}$$

where v is airstream velocity, l is the wing chord (i.e., the distance from leading to trailing edge), and μ is the kinematic viscosity (a measure of internal friction in a fluid) of the air. Reynolds number increases with the velocity of the airstream and with the length of the wing chord. High Reynolds numbers imply turbulence. At Reynolds numbers below 10,000 or so, flying becomes dangerous, for the corresponding boundary layers tend to be laminar and permit separation of the main airstream from the wing. But since the dimensions of a bird's wing are fixed, the only thing it can do to increase Reynolds number, and promote turbulence due to these factors, is to fly faster. Many narrow-winged birds, such as the swift, fly very fast, whereas broad-winged birds like the buzzard can take their time.

Let us summarize briefly a few of the main points of this chapter:

1. An airstream flowing past a wing with convex profile induces a net force on the wing due to "Bernoulli effects." This "lift" is directed perpendicularly to the direction of the airstream (or "relative wind") at any instant.
2. Eddies at the wingtips induce "drag." The pointed tips, and emargination, of bird wings reduce the area over which such eddies can occur.
3. If angle of attack of a wing is increased (leading edge tilted up), drag increases sharply so that forward motion is slowed; danger of stalling arises.
4. If the boundary layer over a wing is turbulent, the airstream appropriate for lift tends to be maintained, even at relatively large angles of attack. Sharp leading edges, turbulence generators on the wing surface, and increased airstream velocity (especially in the case of thick round wings) all tend to promote turbulence in the boundary layer.

Chapter Four

Forms of Bird Flight

Flying without Wingbeats

Earthworms are pitiable creatures. For them, locomotion is an arduous task; slowly but imperturbably they dig their way through the soil. Surely flying animals have an easier time of it. An animal passing through the air encounters considerably less friction than one moving through the ground. Furthermore, flying animals can turn gravity to their advantage; simply by letting themselves fall they can cover a distance of many hundred meters without the slightest effort.

The following example shows how expertly birds have mastered this trick. In the stork colony at Bergenhusen, I had been waiting several hours for the exchange of birds guarding the nest to take place. My arms were already hurting from holding the camera up so long, and my eyes were sore from staring into the sky. At first I did not react at all when a small black dot appeared high above. But when it rapidly grew larger I swept the camera into position–just in time to focus on the stork, which was falling from the sky with its wings furled. Just above the ground it stopped the breathtaking plunge, by stretching out its wings. A few wingbeats provided sufficient braking power, and the stork came to a gentle landing, followed by the obligatory greeting of clattering bills. Shortly afterward, its partner started out. With outstretched wings it glided far out over the meadows on an almost horizontal path.

The two storks provide a striking example of the versatility of flight. Both of them were taking advantage of the gravitational attraction of the earth. The first, flying to the next from a great height, did not need the lifting force of its wings. Therefore, it simply closed them. The other, flying out onto the meadows, spread them far out in order to generate as much lift as possible, so that the fall would be more gradual and the gliding distance extended.

In principle, all organisms can glide, though they do not all do it equally well. Good gliders like the stork cover long distances with very little descent. Animals that approach the earth at a steep angle, on the other hand, are, of course, poor gliders. A quantitative assessment of gliding ability can be made in terms of the glide ratio, the ratio between forward speed (or forward distance traveled in a given time) and sinking speed (or vertical distance traveled in the same time).

For example, if a stork takes off from a nest at a height of 10 meters and lands on the ground after traveling 100 meters, its glide ratio is

Right page: Just above the water surface this rosy pelican can glide especially well. Compression of the air between wings and water increases the lifting force.

100/10, or 10. Buzzards, vultures, and eagles have ratios of about the same magnitude. Albatrosses achieve ratios of about 20, while those of the best-designed gliding aircraft can exceed 40.

The gliding performance of the albatross is accomplished by flying in such a way as to generate the greatest lift and the least drag. There is a certain angle of attack at which these two factors are optimal. Basically, all birds try to set their wings in this optimal range for gliding. The velocity required to do this depends upon the magnitude of the forces pulling downward. This is expressed by the "wing loading," the ratio of the weight of the flying object to the gross wing area. Albatrosses are very heavy and, because of the long but narrow shape of the wings, the wing area is relatively small. The albatross, then, is one of the birds with high wing loading. Therefore, its airspeed must be high if it is to avoid losing altitude rapidly during gliding. (Lift increases as the square of airstream velocity.) Albatrosses can achieve high airspeeds, since their overall drag is small. Their outline as seen from the front is thin, and as they are large birds their friction-generating surface is small with respect to their weight; finally, the elongated, pointed wings generate little induced drag.

Other good gliders, like the buzzard and stork, can glide more slowly because their wing loading is less.

Can all birds with small wing loading glide well? The answer to this must be no. Small songbirds with small aspect ratio, such as sparrows, tits, and blackbirds, are poor gliders even though their wing loading is much less than that of buzzards or storks. Just try to throw a handful of cotton wadding as far as you can; it will fall to the ground within 2 meters, at most. Its mass is so small and its surface-friction drag so great that forward motion is almost immediately halted. Small, round-winged birds are in a similar situation. They have the added problem in that the shape of their wings brings about a marked

A stork landing on the nest. With alulae raised and wings tilted far back, the white stork glides steeply downward, as though hanging from a parachute (upper right). Ruffling of the feathers on the upper surface of the wing (extreme left) and on the underside (third from left) shows that in the last braking wingbeats before landing the airstream separated from the proximal surface of the wing.
From left to right: End of the downstroke, middle of the upstroke, beginning of the downstroke; stopping on the nest; for the take-off the wings are pulled far forward and down.
(Montage of phases taken from several approaches to the nest.)

increase in induced drag. Small birds with pointed wings and slender bodies, such as bee-eaters and swallows, however, can glide quite well.

Very near the surface of the earth or of bodies of water, the glide ratio can be raised by a special effect. The air beneath the flying bird and the surface just below is compressed, so that there is an additional increase in pressure on the underside of the wing and thus in lift. The upward force component becomes greater, and the bird can glide farther.

Poor gliders approach the earth on a steep flight path. This too can be desirable on occasion—for example, when the bird is trying to escape a pursuer. But good gliders are not denied this advantage, for they can also descend steeply when they wish to do so. They need only increase their wing loading, by folding their wings. When a wing is flexed, its center of pressure (the point at which the resultant of the aerodynamic forces is considered to act on the wing) is shifted back, since the area of the wing shifts back. As a result, the centers of pressure of the wings can move behind the bird's center of gravity. This raises the back end of the bird and sets it on a glide path inclined downward. Many birds use this method to enter a steeper glide. The opposite effect is achieved when the wings are spread wide out and forward. Now the upward force on the wings is exerted in front of the center of gravity, and the glide path tilts upward—but only to a certain point. To go higher without beating its wings, the bird must soar.

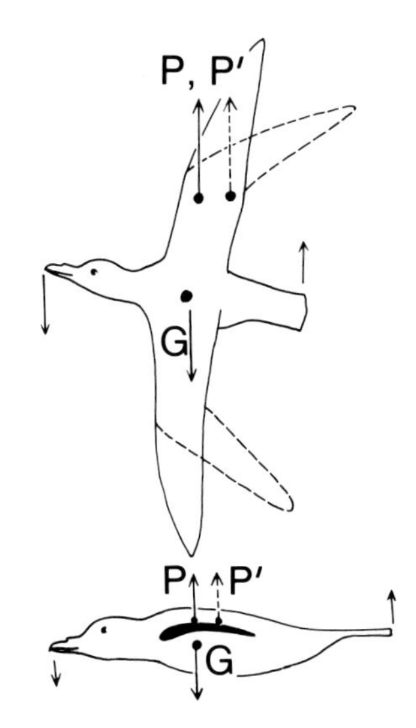

Diagram of the distribution of forces while a downward glide is accelerated by partial flexing of the wings. The centers of pressure (P) move back to P′, behind the center of gravity (G). The upper sketch shows the bird from the side and above; the lower, a view directly from the side.

Long-Distance Soaring

Left page: To glide down more steeply, the fulmar (Fulmarus glacialis) *flexes its wings somewhat (left); to glide upward it stretches them out and forward (right). Above: Downward glide (left) and upward glide (right) of a grey heron just before reaching its landing place.*

When a bird (or sailplane) soars, its altitude is maintained or increased by upward currents of air. It is important that airspeed not be reduced too much during soaring, since sufficient lift must be produced to support the bird. The structural properties that make it difficult for many small birds to glide are even more serious when it comes to soaring; because of their small mass and the large drag they produce, their forward motion is braked too sharply and they quickly lose altitude. Birds that soar well, such as storks, buzzards, and pelicans, keep their descent rate low.

In the autumn, we might well be able to find our stork from Bergenhusen flying with thousands of its conspecifics at an altitude of several hundred meters over the European side of the Bosporus. On their annual migration, the European storks circle over this western tip of Turkey before crossing the straits to Africa. There are no updrafts over the water, but the warmer land generates updrafts strong enough to carry the birds to an altitude from which they can risk the long glide across the sea.

These thermal updrafts (or simply "thermals") are produced when the air near the ground is warmed. The warm air, being lighter than cold air, begins to rise. Sailplane pilots learn to respect the effectiveness of such updrafts—they may be swept several hundred meters upward within a minute. Warming of the air is especially rapid where the sun falls perpendicularly onto a south-facing slope, and where the ground has a high heat capacity (plowed fields, exposed rock, evergreen forests). After a brief warming-up period in the morning, the air over such suitable places begins to rise. It is possible to tell just when this happens by watching the behavior of the birds.

On our travels through the steppes of eastern Africa, we noticed that there were never any vultures wheeling in the air early in the morning. But from about ten o'clock on, the "elevator" was in operation and the vultures began to soar. Seeing them, we knew that

Marabous soaring in a thermal updraft (left) and rosy pelicans soaring in the updraft against a slope (right).

the updrafts had become established. But how did the birds know it? Evidently the sight of only a few circling birds acts as a signal to the others. Then they come flying up from all directions, until several hundred are spiraling higher and higher.

But it often happens that over columnar updraft areas (the so-called dust-devil type of thermal) cumulus clouds form. The warm air over the ground contains a great deal of water vapor. When the air rises, it cools; but cold air cannot hold as much moisture as warm air. The water vapor condenses into droplets, and clouds are formed. Hence, a bird need only fly toward the mound of clouds, visible from a great distance, to be sure of finding an updraft. When a thermal forms only a slender column, a bird unable to fly in tight curves may easily get out of range of the lifting force. For this reason, high-speed gliders like albatrosses cannot exploit such local thermals. Slowly flying, broad-winged birds, which can practically turn in one spot, may, under good conditions, spiral to altitudes of 3 to 5 kilometers.

Thermal soaring is a special aeronautical skill that relatively few birds have mastered. Most soaring birds have large, broad, emarginated wings. When the updraft is very strong, though, birds with smaller and more pointed wings can also be seen soaring—for example, falcons, laughing gulls, and even swifts.

In even more extreme updrafts, soaring can become a sport for everyone. Even auks, with their very high wing loading, can forget about gravity for minutes at a time, without beating their wings or losing altitude. They are pushed upward by the updrafts that arise

along steep coasts. When a prevailing wind encounters such an obstacle, it is necessarily diverted upward. But the small, round-winged songbirds cannot soar even in these obstructional updrafts. They are too light and are soon swept away by the wind. When flying into the wind across an embankment, a lark can remain over one spot in the updraft for a brief moment without beating its wings, but that does not really count as soaring. The masters at soaring are gulls, which can soar in the updrafts over rows of dunes. A common sight at

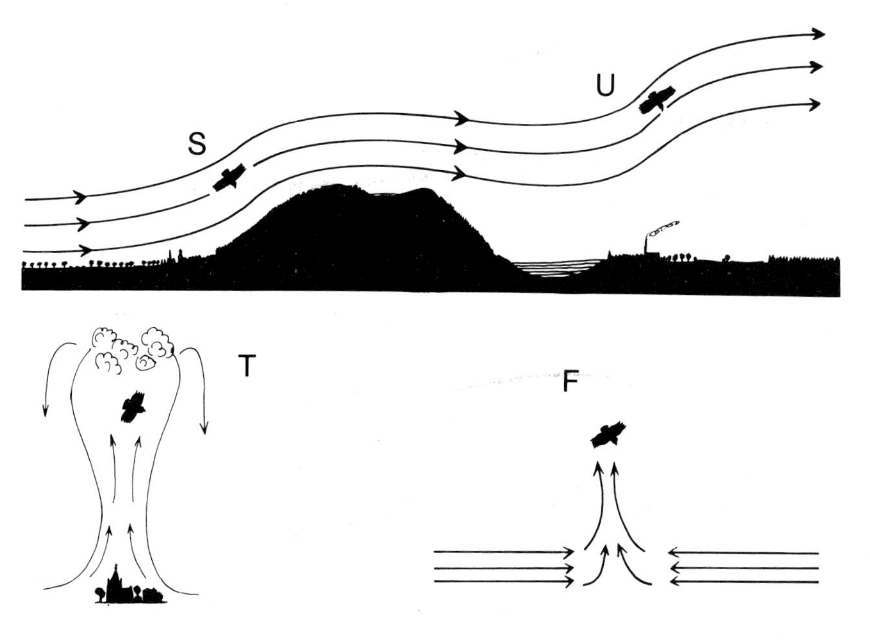

Slope updrafts (S) are produced when wind is deflected over an obstacle, undulating air currents (U) occur several km in the lee of mountains, thermal updrafts (T) are rising warm air, and frontal updrafts (F) occur when two different air masses converge.

A fulmar soaring over the ocean.

sea is a gull soaring motionless in the air behind a moving ship, as though tethered by a string like a kite. Updrafts are produced not only in front of an obstacle; on the lee side the air can be set into wavelike oscillation. This can happen on the downwind sides of mountains, in which case there develop areas of updraft reaching to great heights.

Updrafts, then, can be brought about in many ways. A bird that flies by soaring, and thus depends on updrafts, must be able to detect them. In this respect frigate birds are unexcelled. Over a gradually sloping beach on Tower Island in the Galapagos, where we ourselves would not have suspected an updraft, frigate birds were seen soaring at altitudes between 100 and 400 meters. Slope-produced updrafts could hardly have been felt at that height, but there were certainly other upward-flowing currents of air. Other frigate birds were soaring in the strong uphill wind just off the steep part of the coast. These held their wings slightly flexed. Still other birds were hovering in the updraft produced by undulations of the air on the other side of the island, far leeward of the edge of a cliff. Finally, frigate birds were also soaring directly above the island. These flew forward into the wind and actually gained altitude in the process.

For many birds, slope soaring is the chief method of locomotion. Suitable conditions are found over the ocean as well as over land. At sea, wind is usually blowing, and it is diverted upward somewhat by the slopes of the waves. Stormy petrels and shearwaters soar on the windward sides of the waves. In order not to drift along in the direction the wave is moving, from time to time they jump over to the next approaching wave. There are no thermal updrafts over the ocean, but there is one other method of soaring that is independent of the waves: dynamic soaring. Albatrosses show us how this is done.

The Secret of Albatross Flight

A horizontal gust of wind striking a flying bird from the front generates lift, because of the increase in airstream velocity. The bird is swept upward. If such gusts arrived in regular succession, the bird could use them to cover long distances without beating its wings. In theory, the sequence of events would be as follows. When the bird is pushed upward into the wind it would lose speed but gain altitude. During the subsequent downward glide its flight speed could increase. A new gust of wind arriving at just the right time would then carry the bird upward until the speed it had gained was lost again.

Birds certainly use the method just described to make fairly long flights, when they happen to encounter such winds, but the principle is not very useful for really prolonged soaring. Gusts of wind, after all, are not regular occurrences, and sometimes they do not occur at all. But we do find a completely systematic rise in wind velocity over the ocean, owing to the fact that wind is slowed by friction at the surface of the water. The closer to the water, the more the wind is slowed; wind speed rises with altitude. Albatrosses take advantage of this wind gradient. They rise into the wind, thus encountering steadily increasing wind speed. The increasing lift drives the bird still further up until it has lost momentum; then it glides down at an angle to the wind, in

The albatross technique of dynamic soaring. The progressive lengthening of the higher arrows represents increasing wind speed.

order to build up speed. Having done so, it once again rises into the wind and repeats the entire process. This is the principle of "dynamic soaring," discovered by the Frenchman Idrac 40 years ago during his long expeditions in the South Atlantic. The albatross combines this method of soaring with that of using the updrafts at the slopes of the waves.

In the Galapagos, on the steep coastline of Hood Island, we were fascinated by the sight of these magnificent large birds, sailing past us almost close enough to touch in the glow of the setting sun. They looked like little sailplanes, except that their maneuvers were much more rapid and intricate; with no suggestion of uncertainty, one swooping, curving maneuver follows another.

Gliding and Soaring on a Curved Path

In an airplane, the pilot manipulates a variety of controls—elevators, ailerons, and rudder. The causes and effects involved in directional changes are easy to follow. When watching a flying bird, we detect only the effects of its control operations, while the actual mechanisms are almost always concealed. Usually the best we can do is to discover which of the many possible modes of control is the most likely.

Birds can raise or lower the wings, move them forward or back, reduce the wing area, rotate the wings at the shoulder to produce different angles of attack, or warp (twist) the wings; and during all this they can move the tail as well. All these steering operations can be graded in extent and combined in various ways. The resulting

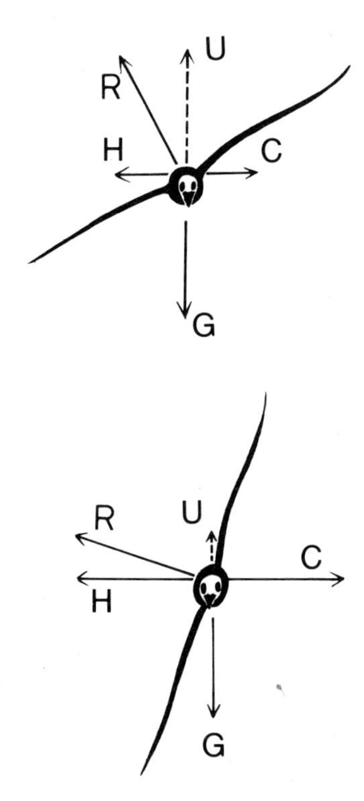

Diagram of the distribution of forces on a bird banking in a curve. C, centrifugal force; H, horizontal component of the resultant force (R). The resultant force, generated on the convex upper surface of the wing, is directed inward when the bird banks. In a sharp curve (below) C and H are larger, and the upward component (U) decreases. G = gravity.

diversity of ways to achieve curved flight is large indeed.

Just as a bicyclist leans in the direction he wants to turn, aircraft and birds bank when flying in a curve. If they did not, centrifugal force would pull them out of the curve. The smaller the radius of curvature and the higher the flight speed, the larger is the centrifugal force. To compensate for the increased centrifugal force in a tight curve, the cyclist leans and the bird tilts at a steeper angle. The cyclist, so to speak, lets himself fall into the curve

Airplanes and birds use the lift generated by the wings to counteract centrifugal force. They bank in a direction such that the upper surfaces of their wings, and thus the lift produced there, have a component toward the center of curvature. When their inclination is great enough so that the horizontal, centripetal component of the resultant force is exactly as large as the centrifugal force, a stable position for curved flight has been reached. Under greater centrifugal force the bird must bank more sharply; with a reduction in centrifugal force, it can return toward the horizontal. When a bird flies in a very tight curve, its inner wingtip can point almost straight down. In this position the resultant force is almost horizontal, and its vertical component is very small. In order not to lose too much altitude in such a turn, the bird must build up speed beforehand, so as to increase lift.

How do rapidly gliding birds bank by the necessary amount? On several occasions I have seen fulmars change their inclination by downward movements of the wing on the inside of the curve. But I could not tell whether the wing was pulled down by contraction of the wing musculature or by forward warping of the hand part, with the consequent sudden loss of lift for that wing. In any case, it was clear that following this movement the bird was in an inclined position. There are sure to be many possible ways of achieving the same thing. Here, too, several operations are probably combined.

Similarly, birds have different ways of flying in slow curves. The most effective method is certainly that of braking on one side (as people do in sledding or canoeing). The disadvantage here is that it entails loss of forward speed. A soaring bird can ill afford a loss of momentum, and a gliding bird can afford it only if speed is recovered later by steep downward flight. I have often seen fulmars, gannets, and buzzards initiate a gradual curve by unilateral braking, with the wing on one side set at a steep angle of attack. Frequently this unilateral wing tilting is followed by partial flexing of the same wing, which reduces lift. In this way the bird banks by the amount necessary for the curve.

Banking in a curve and unilateral braking can also be achieved by setting the wing at a negative angle (pitched downward) to the flight

On the following pages are shown various techniques for flying along a curve: Banking by the herring gull (p. 74) and by the black kite (Milvus migrans, *p. 75, bottom); a fulmar with the angle of attack of the right wing increased (p. 75, middle); a hooded vulture pulling the right wing, set at a large angle of attack, far forward (p. 75, top).*

path and flexing it at the same time. Gannets habitually use this method to make sharp curves while flying slowly; the maneuver is followed by a sudden downward plunge. During these slow curves, banking serves less to counteract centrifugal force than to put the bird in an unstable position from which it can descend and quickly build up speed.

The lower the flight speed in a curve, the smaller the effect of centrifugal force. A slowly gliding buzzard could turn simply by slightly reducing the area of the inside wing, without banking at all.

Birds soaring in thermal updrafts must fly with wings and tail fully outspread. In this case, the problem is to generate as much lift as possible. What turning techniques are feasible for such birds?

If plenty of lift is available, they can expend energy relatively freely and acquire the torque necessary for turning by braking on one side—for example, by increasing the angle of attack of the wing on the inside of the curve. In so doing, though, they must be careful that the maneuver does not induce bank in the wrong direction, since the wing with the larger angle of attack generates more lift. To compensate for this, the bird can, for example, reduce the area of that wing by flexing it. What does this operation achieve? The result is lowering of the wing (since its loading increases) and an even greater braking effect (since the induced drag increases because of the smaller aspect ratio). The "correct" bank angle for the curve would thus be assured.

On the other hand, in a weaker updraft the bird must ration its energy; angle of attack, extension, and warp of the wings must be changed only slightly so as not to reduce momentum.

The tail also plays a role in curving flight. It can generate torque if it is held at an angle or lowered on one side. Such torque may either help to establish the desired curve or correct inappropriate turning movements of the bird.

If you have ever watched the steering maneuvers of a gliding or soaring bird, you will have seen how difficult it is to discern the cause of each change of direction. Slight warping, tilting, or other changes of wing setting can rarely be followed exactly because of the distortion due to perspective. Furthermore, steering maneuvers can be evaluated properly only if the air currents at the site of the bird are taken into account, and this information is largely inaccessible to an observer on the ground.

Stabilization of Gliding and Soaring Flight

At any time there may be a sudden change in the direction or speed of the wind. These do not deflect a bird from its course, even though it may be in the middle of a complicated steering maneuver. Only when struck by a strong gust does the bird take countermeasures. By their very structure, birds are protected from forces tending to change flight attitude.

Almost all birds are built like high-wing monoplanes; the body hangs from the wings like a pendulum, so that it tends automatically to return to the normal position. Moreover, certain birds, including those that make slow hunting flights (like the kites as well as gliding pigeons and tree pipits), hold the wings up in a V-shaped posture. This "positive dihedral" configuration also has a stabilizing effect during landing on water.

How is this stabilizing effect produced? If a bird holding its wings in this way tilts to the left or right, the surface of the lower wing generates more vertical lift than that of the upper wing, and the bird returns automatically to the horizontal. It is true that the lift produced by each of the wings is the same, but when a wing is inclined, upward lift, being perpendicular to the wing surface, is not vertical. The vertical component of the aerodynamic forces is therefore considerably smaller in the upper wing than in the lower; the surface of the lower wing is almost horizontal, and lift there is exerted almost straight upward.

Birds with wings held to form a deep V are very stable in flight. Sailplanes similarly constructed are "well-behaved" designs; when tilted to one side they immediately right themselves. Sailplanes for stunt fliers are built differently. Their wings are not V-shaped, but straight. Stunt fliers desire a certain amount of instability. Such a plane can more easily tilt suddenly to one side and plunge downward—an ability often used to start out a breathtaking aerial maneuver.

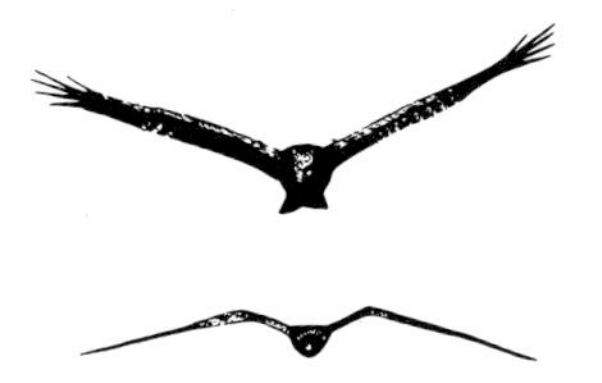

A positive dihedral (e.g., the kite, above) provides stable orientation in flight, while a negative dihedral (e.g., the frigate bird, below) leads to instability.

Frigate birds have also evolved this unstable design. They hurtle instantly down from a soaring flight to catch their prey. In this case a positive dihedral would be no advantage because, with greater stability it would take the birds longer to start a new maneuver. The outline of a frigate bird, seen from the front, is like that of a highly maneuverable sailplane except that the inverted V is interrupted by a slight bend in the wings near the base.

Besides the low center of gravity of the bird body and the dihedral configuration of the wings, the wing outline as seen from above is an important factor in automatic flight stabilization. Airplane wings, for example, may be swept back; those with sweepback of 45° or more are called arrowhead wings. Birds can give each of their wings such an attitude by flexing them slightly, so that the tips aim back and the wrist joint forms the point of the arrow. Airplanes so constructed tend to be inherently stable, returning eventually to horizontal flight, despite the sideslip associated with complicated roll and yaw maneuvers. That is, during sideslip, the wing that is angled more perpendicularly with respect to the airstream exhibits greater lift than the other, which is necessarily angled back even more. In slowly soaring birds of prey, the tail is in constant motion and acts as a supplementary stabilizing device.

If the mechanisms of steering and stabilization in gliding and soaring flight are difficult to work out, flapping flight will certainly confront us with even greater puzzles.

Flapping Flight—Nature's Prerogative

"On the wings of the wind" is not just a bit of poetry; a bird's own wings would be of little use if it were not for the wind flowing over them. But what happens if there is no wind, or if a downward glide path is not desired? The bird, or at least its wings, must of course move with respect to the air and maintain this motion.

When taking off or hovering in one place, birds beat their wings forward and down, so that the air flows around them. Once the bird is moving forward through the air, that motion itself of course, produces a relative wind.

The most important element in the downward wingstroke is the twisting of the wing toward its tip in such a way that the angle of attack gradually becomes smaller. As a result, both direction and magnitude of the lift generated by the wing change continuously from base to tip. Near the body, in the arm section, the wing moves over a relatively short distance as it beats. Here the airstream is composed mainly of the flight-induced headwind, with only a small component from the wingstroke itself. The consequence is that lift, being perpendicular to the nearly horizontal airstream, is directed upward. The lift on the arm part of the wing bears the weight of the bird. At the wingtip the situation is different. Here the airstream is generated chiefly by the beating of the wing; the wingtip, being further away from the body, moves over a much greater distance per beat, and hence at a much higher speed. The lift produced here is directed more forward (and is larger) because of the forward rotation during the downstroke. The manus thus provides thrust and can be thought of as the propeller.

Because of the twist in the wing, a bird holding its wings at a large angle of attack (during braking, for example) can still generate lift near the end of the wing, where the angle is smallest.

When a bar-tailed godwit (Limosa lapponica, *above), or crowned crane* (Balearica pavonina, *below) takes off, the wings are strongly twisted and the stroke carried far forward. Because of the twist, the upper surface of the wing is visible near the base and the lower surface, near the tip. The sketch shows the distribution of forces in flapping flight: AA, direction of airstream over the arm; AM, airstream over the manus (this is larger and comes more from below); L, lift; D, drag; R, resultant force; T, thrust component.*

R
L
W
AA
W
R
L
T
AM

Rapid Flight by Large Birds

Large, heavy birds of course feel the effects of gravity more than lighter ones. To counteract this, large birds must produce a great deal of lift. The second hindrance to flight, drag, is also greater in these large birds during high-speed flight, but their larger mass is also decelerated less by it.

Since most large birds beat their wings relatively slowly, they must keep themselves in the air or gain altitude by generating lift in both the upstroke and the downstroke. They manage this as follows. During the upstroke the proximal part of the wing is extended, generating an upward force to bear the bird's weight. The wing is raised with the leading edge upward, so that it is tilted at a considerable angle to the airstream associated with forward motion of the bird. As a result, lift production is accompanied by production of a backward-directed force. This by-product of the motion is much smaller than the thrust produced in the downstroke and does not produce great deceleration because of the large inertia of the moving body of the bird.

Most of the lift, as well as thrust, is generated during the downstroke. The effectiveness of this wing movement can readily be observed when large birds, such as swans, geese, or eagles, are taking off. At every wingstroke the body is jerked upward. But, if the bird has a long neck, the head does not follow this upward movement immediately. It is automatically held in its prestroke position with respect to the surroundings. This complicated reflex, which uses eye and neck muscles to hold the visual scene fixed during parts of locomotor movements, is a very common one and has clear visual

Upstroke of a short-eared owl soon after take-off. The manus moves up with the anterior edge leading (less drag).

advantages. For example, watch the head of a walking pigeon during each forward step.

Squacco herons (Ardeola ralloides) *taking off. At left the beginning of the downstroke, and at right the end. In the rapid flight that follows, the wings are no longer pulled so far forward; the stroke tends more to be straight down, for once the bird has reached cruising speed only the force necessary to overcome drag is required.*

Slow Flight by Large Birds

When large birds fly more slowly, the airstream associated with forward movement becomes weaker. The bird must take extra steps to compensate for the loss of lift. In order to increase the flow of air past the wings, the bird beats them more rapidly and through a larger angle. But this alone is often not enough to generate sufficient lift. In certain situations–during vertical ascents, when altitude must be gained with no headwind, and during descents, when as much of the bird's momentum as possible should be dissipated in the air–large birds like herons or birds of prey and medium-sized birds like parrots or ducks use a special method. They rotate the manus through almost 180° during the upstroke, bending it backward. In the process, the primary feathers spread apart so that the air can pass between them. In this position, the airstream over the manus moves from back to front. Each individual feather can, in this situation, generate a lifting force. We shall encounter this flight with rotated manus again, in discussing landing and take-off procedures (p. 118 and illustrations on pp. 83, 84 and 120).

Three phases in slow flight of a falcon. Bottom right: upstroke with manus inverted and primaries splayed. Top: downstroke with strong twisting, alulae raised, and covert feathers in the proximal part ruffled. Left: in the closing phase of the downstroke the wings are pulled far forward (a large-amplitude stroke increases the air flow) and the alulae are raised.

In a slow flight downward, the pigeon must compensate for the lack of headwind by larger-amplitude wing strokes. Upper left: beginning of the downstroke. Upper right: end of the downstroke. Bottom: upstroke with manus inverted.

Small Birds Fly Differently

We have mentioned several times that small birds, such as sparrows, finches, and blackbirds, do not find it so difficult to stay in the air; for them, the great problem is moving forward. Their mass is so small that drag has disproportionately large effects. To overcome this braking effect, small songbirds must generate as much thrust as possible. On closer inspection, the wings of small birds in operation look like real propellers. To see this, however, one needs greatly enlarged photographs made with 1/40,000-second flashes. Comparison of the positions of the wings at different instants during the stroke reveals the aerodynamically significant changes in wing and feathers. The feathers of small birds are more delicate than those of large birds. But the airstream speeds produced during the rapid wingstroke are not lower than in large birds—in fact, they tend to be higher. Those single primaries separated from the smooth layer of feathers are bent far back by such airstreams. Moreover, their wide outer vanes fold back and up during the stroke.

In the downstroke (B and C) the wing of a small bird changes shape (A, resting). In the upstroke (right) the primaries separate, letting the air through, so that little drag is produced; here this is demonstrated by a chaffinch (Fringilla coelebs, *view from behind).*

What might be the consequences of such structural changes? Folding back the outer vanes amounts to reduction of the angle of attack. Thus, the forces generated by these feathers would be directed more forward and would act to pull the bird forward. The angles of

attack would be reduced not only by the passive warping of the primary feathers but also by a deflection of the airstream. Just as a wing turns the air striking it downward, so each feather would divert the air flowing past it. Each feather standing apart from the others changes the direction of the airstream encountered by the next feather; each feather directs the air toward the following feather. As a result, the feathers "downwind" of the leading feathers have an effectively smaller angle of attack.

To appreciate what happens during the exceedingly rapid stroke of a small bird's wing, one needs something more than single "stop-motion" photographs. Slow-motion replays of moving-picture films make visible the details of the movements. In making such films, the exposures must be repeated at a rate of several thousand frames per second. When these films are projected at normal speed (24 frames per second), the movements are enormously slowed. In using the method to study free-flying redstarts, we have slowed their motion by a factor of about 70.

A Study in Slow Motion—the Flight of a Small Bird

There was an atmosphere of tension in the darkened tent that concealed us; the only thing visible from outside was the lens of the high-speed camera with which we shared the tent. The spotlights mounted around the redstarts' brood box were still turned off. Several days ago we had put together the large portable tent to serve as camouflage, and every day since then we had moved it a little closer to the birds' box. The box, with the baby birds in it, at first was hanging on a wall; each day we had put it a little lower on the wall, and finally we had moved it to a portable stand. One by one we set up the spotlights. Our final step was to set a light-colored background of cardboard behind and to the side of the box so that the birds showed up clearly. Whenever the parents appeared with food in their beaks, we discreetly withdrew. What with our cautious behavior, and with the added inducement of a feeding dish full of mealworms, our redstarts soon lost all trace of shyness. They continued to feed their young even when the five spotlights were bathing the scene in glaring light. But wouldn't they be too startled when the camera started to run? We sat and waited. One of us had a finger on the switch to turn on the light, another held between thumb and forefinger a cable with which he could set the camera in motion, and the third kept watch over the nest box, a piece of string in his hand. This string was of the utmost importance. Its other end was attached to a cover over the opening of the box. When a redstart flew up, the observer in the tent could operate the string and instantly open or close the box.

Using a great tit (Parus major) *as an example, the photos show the downstroke typical of round-winged small birds (in sequence from left to right and top to bottom): the wings, fully extended, are drawn downward diagonally, then folded against the body, and pulled up into the starting position.*

A shadow hopped nearer through the shrubbery. "Light on!" It was the female. She started toward the box. "OK!" The camera began to whine. One second, two seconds... "That's it!" The film (thousands of frames!) was finished. The bird had not been bothered by the noise, but it had been brought to a halt by the falling lid of the box. It stopped in midair, hovered a moment, flew backwards, and then forward when the lid was opened. We breathed a sigh of relief—the experiment had worked. From that point on, film was used up at a greater rate than I had ever experienced. The box was turned, and the birds were photographed in different ways—from behind, *via* a mirror, or from two directions at once.

Our satisfaction turned to delight when, weeks later (the young redstarts had long since left the nest), we looked at the developed films. Sequences of movement never before seen, and impossible to see without high-speed photography, were captured there. Each scene revealed something new; each maneuver was a lesson in cause and effect.

In One Spot—Forward—Backward

Redstarts can hover in one spot, even in still air, for 10 or 20 seconds—longer than most other small birds. To do this, they beat their fully extended wings downward at a 45-degree angle. (A wingstroke can best be described by the path followed by the tip of the wing.) At the end of the downstroke, first the arm and then the manus of the wing are flexed. Completely furled, the wing is pressed to the body. The primary feathers swing past the body in such a way that they can be drawn upward with the anterior edge leading. In the process, they separate so that air is let through the gaps between them. This considerably reduces drag, just as it is less for a tennis racket than for a fan. Then, without a pause, the wings are stretched upward and again brought down following an elliptical (as seen from the side) path. At the end of each downstroke the birds pull the wings forward and up. To prevent the wing movements from turning the body, the tail is spread wide after each downstroke. Although the sequence of movements appears fairly uniform, the speed of the wingbeat can vary widely in different maneuvers. On one hand, the beat frequency can range from 14 to 20 beats per second, and on the other the ratio of the durations of upstroke and downstroke can change (ratios from 0.8:1 to 1.5:1 have been measured). During the power-producing downstroke there is a continual change in direction of the airstream, since the wings are moving in a different direction at each point on the path.

Path of the wingtip of a hovering redstart (Phoenicurus phoenicurus)*; side view on the left, view from below on the right.*

When a redstart is hovering, its wings generate a net average lifting force directed straight upward, which must of course be exactly as large as the downward pull of the weight of the body. This net force is the average of all the constantly changing aerodynamic forces acting on the wing. There must be an equal and opposite force, and this takes the form of a downward acceleration of air.

A simple experiment makes this "downwash" visible. A tabletop is covered with down feathers. I set my tame sparrow on my index finger and hold it over the table. Then I jerk my finger down, out of its grasp. For a second, the startled sparrow holds itself in place by beating its wings. As it does so, the down on the table below is blown away by the blast of air; it flies equally in all directions and piles up in a ring around the bird's position. To demonstrate the direction of the backwash produced in normal forward flight, I replace the sparrow on my finger and, having again spread out the layer of down, hold the bird over the table until it flies off of its own accord. Now the feathers are swept away in a direction just opposite to the direction of flight. It is easy to imagine that the wing movements producing this backwash must have been different from those in hovering flight. At the transition to forward flight the angle of the downstroke changes from 45° below horizontal to a much steeper angle, almost 80°. Now the airstream comes almost directly from below the bird; the lift is directed toward the front and pulls the bird forward. At the transition to forward flight small birds raise their wingbeat frequency by about 20%, in order to increase the speed of the airstream. The feathers at the end of the wing are bent back even more strongly and the outer vanes fold further upward. Accordingly, the angle of attack of these feathers is reduced. There is an additional active reduction of the angle, brought about by forward tilting of the wing. The tail is bent

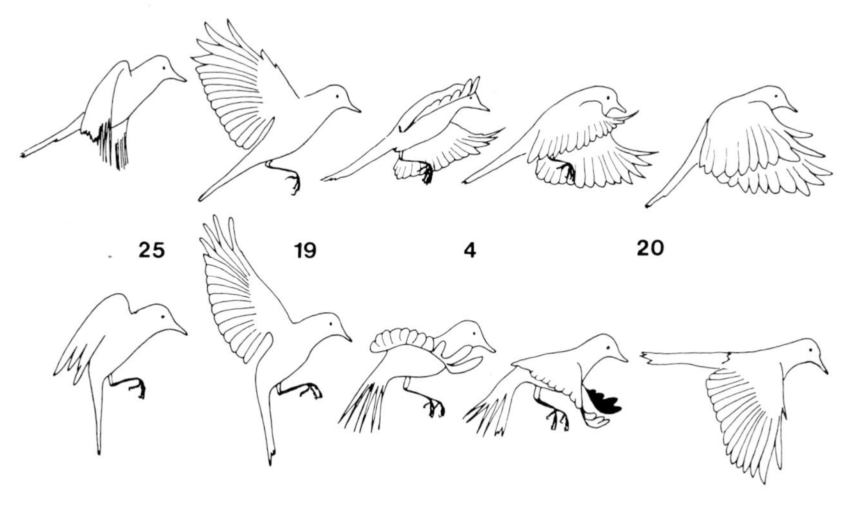

In the transition from hovering (upper row) to straight forward (lower row) flight the wings are drawn more directly down. The drawings were made from frames of a film; the numbers indicate how many frames separated those selected (exposure rate, 1600 frames per second).

downward to increase the tendency of the body to pitch forward, so that it comes into a horizontal orientation for forward flight.

Once forward flight has begun, the airstream over the wings no longer comes exclusively from below; the bird's own passage through the air creates a stream from front to back. The average airstream, then, most likely comes from a direction ahead of and below the bird's body.

Redstarts can even fly backward, though not for very long. In backward flight the wings beat with the undersides facing away from the direction of flight. In the proximal part of the wing the aerodynamic angle of attack is over 70°, and in the manus I estimate it as about 20°. The tips of the outer primaries are passively twisted to a still shallower angle. It can be assumed that in backward flight the airstream separates from the wing over most of its surface. With such large angles of attack, none of the techniques available to help the bird in high-lift situations can prevent disruption of the airstream—neither the roughness of the wing surface, a potential turbulence generator, nor the raising of the alula to form a "leading-edge slot." Perhaps the rhythmical beating of the wings prevents separation of the airstream. This seems possible, since if a wing is moved back and

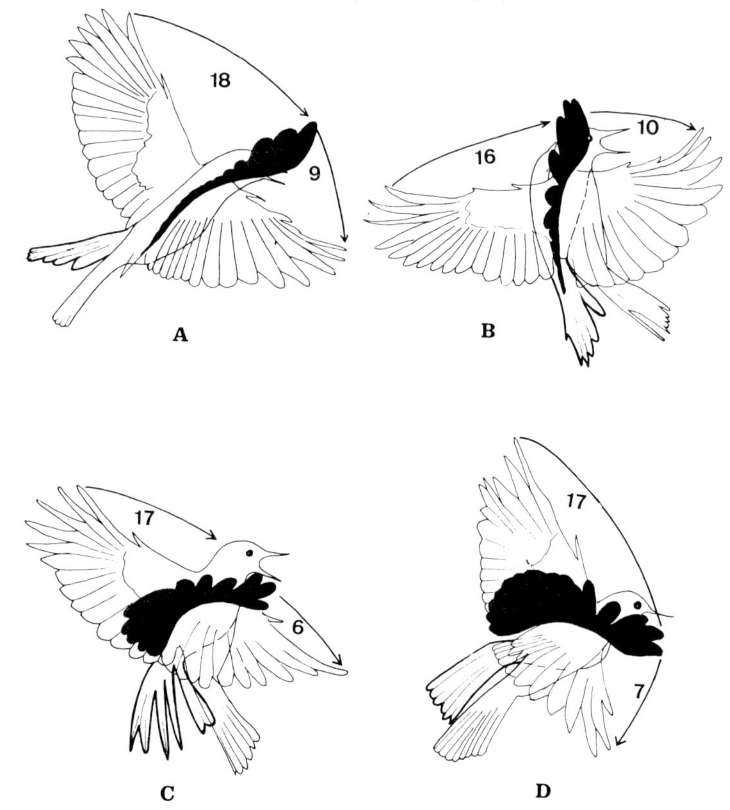

Hovering flight (A), backward flight (B), transition to forward flight (C), and beginning of forward flight (D) of the redstart. The arrows indicate the direction of the diwn-stroke, and the numbers show the stroke duration (time between the wing positions in milliseconds). Each sketch shows three positions of the right wing: beginning, middle (black = underside of wing), and end of the downstroke.

forth in a current of air, the airstream adheres to the wing because of inertia even at large angles of attack. In order to test whether the airstream actually does break free at angles as large as those in backward flight, we did some experiments in a wind tunnel. Thin, very flexible barbs from goose-down feathers were glued to redstart wings mounted in the airstream so that the angle of attack could be varied. At which angles would the down barbs lie against the wing in the direction of the stream, and at which would they be whirled around?

Our indicators showed that the airstream separated when the angle had been raised to only 35-40°. This would imply that during backward flight the stream has separated from most of the wing surface. Unfortunately, these wind-tunnel experiments provide such a poor imitation of the air currents in the real situation that the results do not really tell us so much. For one thing, the airstream around the wing of a free-flying bird does not have the same velocity from the tip to the base of the wing. The air flows fastest at the tip, since the end of the wing travels through a much longer distance in a given time than does the part near the base. And there are other differences. Isolated wings are stiffer; since they cannot be so easily distorted as the wings on a living bird, their interactions with air currents are different. Moreover, the experiment did not simulate the active changes in shape of the wing during a stroke. For these reasons, we glued fine bits of down to the wings of free-flying birds in the hope of photographing them. But when the redstarts next flew in front of the camera, they had neatly picked off all our little indicators!

From the data provided by research on the behavior of airstreams, as well as from the external appearance of the downstroke during backward flight (the wing does not move smoothly forward, but flutters during the movement), it can be concluded that the birds primarily shovel air forward in order to move themselves back. That the wings generate considerable drag is also indicated by the wingbeat frequency, which is lower than in the other maneuvers.

But why does the bird not fall down, if it is producing so much drag and little lift? There are several reasons. First, small birds are very light and so tend not to fall very fast (but can, by the same token, be accelerated very quickly). Second, the primary feathers are so delicate, and therefore so strongly bent during the downstroke, that they assume a relatively small angle of attack, and even then continue to generate force components contributing to lift. Both of these properties are important in backward flight, and one will seek them in vain in larger birds. Large birds can at most make a short jump backward through the air.

And there is another area in which large birds are excelled by the small ones–flying in tight curves.

Turning On a Dime

The redstart can use either of its wings to generate force directed either forward or backward. The way this is done has been described in the sections on straight-forward and backward flight. Now, if the bird beats one of its wings as in forward flight and the other as in

The technique small birds use to fly in a curve: the redstart first curves to the left (first and second phases from the top) and then to the right. The wing on the outside of the curve moves more steeply downward than the other, which is set at a large angle of attack. (Phases taken from a film sequence exposed at 1600 frames per second.)

backward flight, it will rotate. With a single wingbeat, lasting only 0.05 second, it can turn itself through almost 180°. In these almost instantaneous pirouettes the wings are pulled down more rapidly than in hovering flight. The wingbeat frequency is increased from less than 20 beats per second to 24 beats, and the ratio of the durations of upstroke and downstroke can fall from *ca.* 1:1 to 0.4:1, so that the ineffective upstroke occupies less time. We can be certain that the

Right page: a white stork curving left (above) with steep downstrokes of the outside (right) wing. The inside wing is set at a larger angle of attack, as is the inner (now right) wing of the rightward-curving short-eared owl (middle, downstroke; bottom, upstroke).

increase in the wingbeat frequency offers two advantages: the direction of the wingstroke, and thus the flight direction, can be changed sooner, and since the strokes follow in more rapid succession, more lift is generated in a given time. For fast turns additional measures are sometimes taken, such as carrying the stroke of one wing, that on the outside of the curved path, far down and ahead of the body.

Extreme flight maneuvers such as instant on-the-spot turns expend a lot of energy. One can see this from the way such birds rest as often as they can, usually before flight maneuvers. They do this in brief pauses, during which the wings are simply kept still for a moment while they are held close to the body during the upstroke. The moment of stillness may serve not only to refresh the muscles but to allow better visual orientation prior to a new flight maneuver.

Many small birds routinely make pauses between bursts of wingbeats; a graphic name for this kind of flight is "trill flight." The alternation of exertion and rest produces an undulating flight path. Tits and sparrows fly this way, and so do woodpeckers and jays. In this case, too, the pauses can be assumed to allow recovery of the muscular metabolism or visual orientation.

The high wingbeat frequency, together with the low mass and the relatively large amount of work done by the muscles, enables small songbirds to start, slow down, and change direction extremely quickly. They are outperformed in such maneuvers only by still smaller birds—the hummingbirds.

Hummingbirds Fly More Like Insects

The wings of hummingbirds are about the same size as those of large insects. They beat at a similar rate and move over a similar path, rather like a signaler waving a flag in the form of a figure eight. The smaller a hummingbird is, the more rapidly it beats its wings—up to 100 times per second. The way hummingbird wings operate can best be compared with the action of a propeller. But whereas the propeller rotates in one direction, the hummingbird beats its wings back and forth in such a way that the anterior edge always leads; thus the airstream is always from front to back of the wing, although in the downstroke it comes from the front, and in the upstroke from the back, of the bird. Since the wings are passively bent convex upward, lift is generated during both strokes of the cycle. The thrust produced by a propeller, which draws an airplane forward, is directed upward in a hummingbird—as it is in a helicopter. Erich von Holst described this well in saying that the bird "hangs from its propeller." Hummingbirds

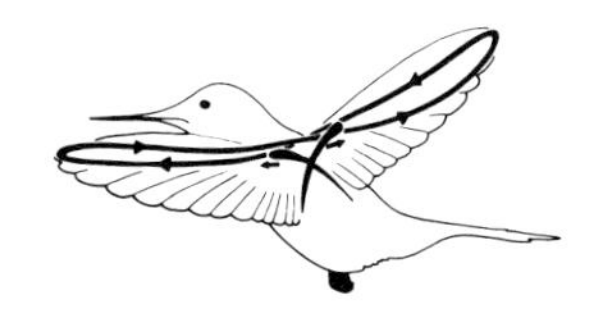

The path of the left wingtip of a hovering hummingbird is a flattened, almost horizontal figure eight. The wing is convex-upward to an equal extent during both upstroke and downstroke, as the profiles drawn in the middle of the body show.

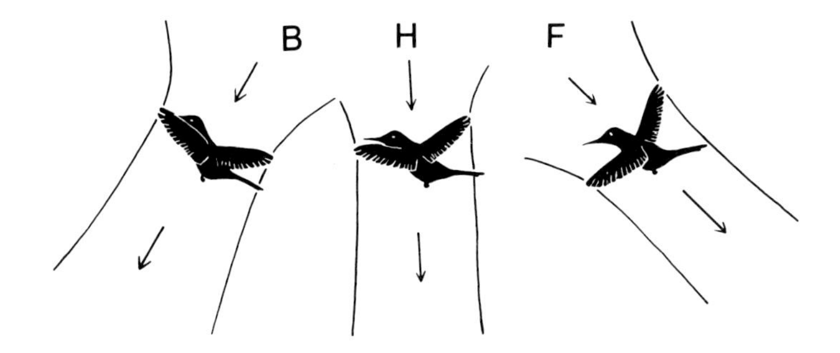

Hummingbirds change their direction of flight by tilting the plane of the wingstroke. The backwash from the wings changes angle accordingly. B, backward flight; H, hovering; F, forward flight. The arrows show the direction of the backwash and the lines at either side, its approximate extent.

produce a downwash just as the rotors of a helicopter do. The effect is like that demonstrated in the sparrow-and-down experiment. When the wash aims straight down, the hummingbird hovers in one spot. If the bird changes the direction of its wingbeat so as to make the figure eight tilt down in front, the wash is aimed toward the back and the bird moves forward. Conversely, a hummingbird flies backward when the plane of the beat is oppositely tilted, so that the front end is raised and the blast of air aims down and forward. To move to one side, the bird beats the wing on the side away from the movement at a higher average angle; the force it generates is then directed appropriately.

Hummingbirds, too, vary the wingbeat frequency in different flight maneuvers. During lateral changes of direction by the species I studied, *Chlorostilbon melanorhynchus,* the range was 42-74 strokes per second. During hovering, on the other hand, the variation is less. Here, too, it is a general rule that in extreme maneuvers the duration of the wingstroke is shortened. The ratio of upstroke and downstroke durations also changes.

Anyone who observes flying hummingbirds will need no complicated measurements to recognize immediately that they are the most maneuverable of all fliers. In the blink of an eye, like overgrown hoverflies, they flit from flower to flower. With extreme suddenness they stop in midflight, only to rocket away again. The smaller the species of hummingbird, the more quickly these changes are carried out. The smaller of two hummingbirds in my flight cage was always the more agile when pursued. It led the larger bird a merry dance, moving out of reach as it liked.

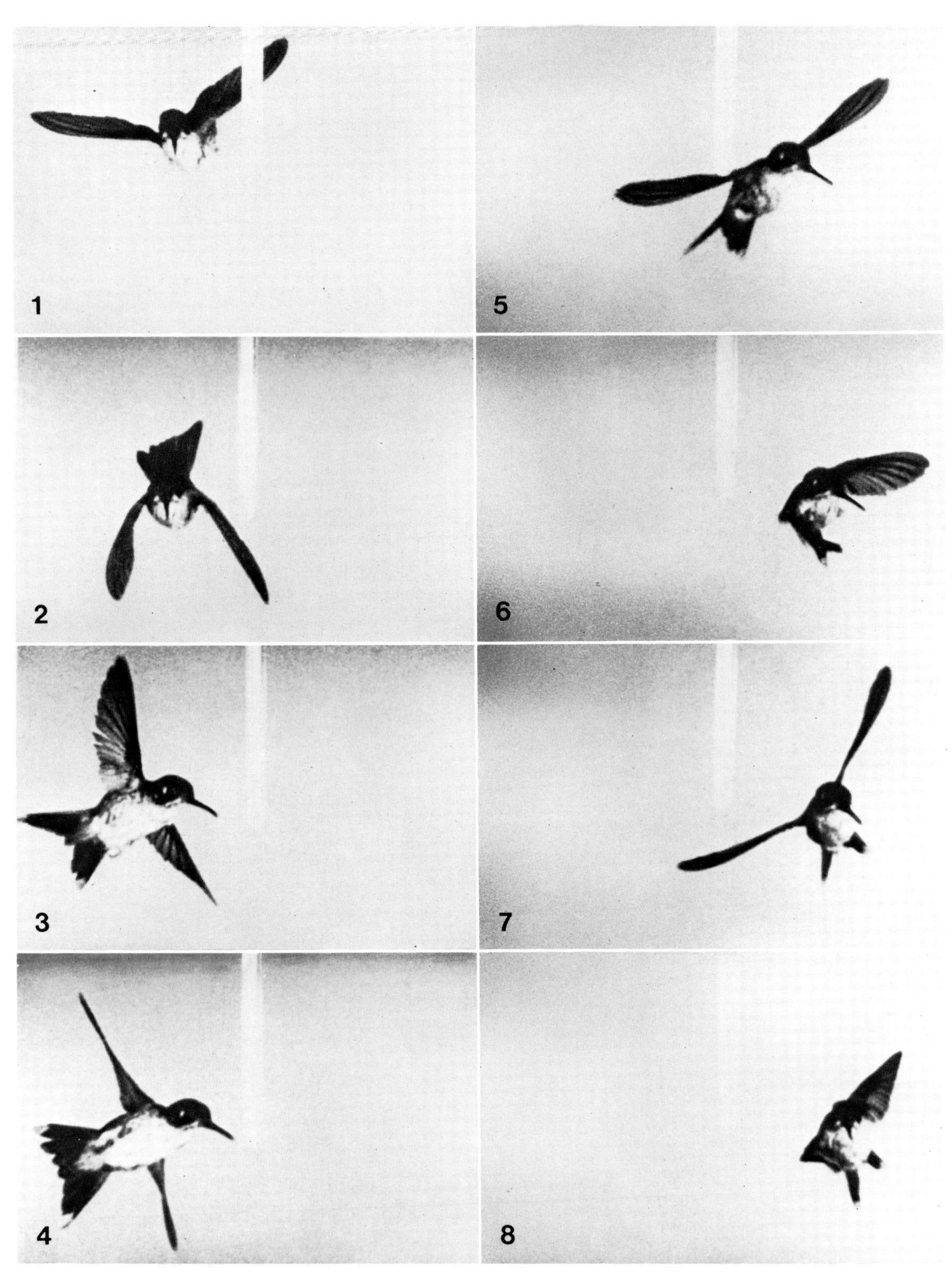

Phases of hummingbird flight, taken from a film exposed at 3000 frames per second. In a curve to the left (3 and 4) the right wing beats higher; in the subsequent rightward curve (5-8) the left wing beats higher.

Characteristic phases of a wingbeat during hovering (viewed from above). Like someone waving a signal flag, the bird turns its wings 180° at the end of the downstroke (2) and the end of the upstroke (6).

CHAPTER FIVE

The Maneuvers in Which Birds Excel

Take-Off and Landing

"Plane Crashes in Attempt to Land... Stall During Take-Off Leads to Disaster." Headlines like these make the front pages only too often.

Most airplane crashes occur on take-off or landing. To take off successfully a plane must accelerate rapidly to a high speed; the wings must be able to generate enough lift to support the huge steel body. But the ascent must not be too steep, or the airstream will break away from the wings. In landing, on the other hand, speed must be reduced so precisely that the minimum gliding speed is reached no earlier than the moment the plane touches down on the runway. Not until then can full braking power be applied. Fog, ice, and severe storms can rule out take-offs and landings altogether.

Human efforts to fly are repeatedly frustrated by these limitations—difficulties and dangers birds never have to face.

Birds take off and land in any weather, even at night, without headlights or ground-control systems. A bird's landing on a waving branch is precise to the last millimeter, and it can take off straight upward.

Why do birds fly more safely than airplanes? A simple formula suffices to answer this: birds are light and fly slowly, whereas airplanes are heavy and fly very fast. As we know, it is difficult to accelerate a heavy body, and almost as difficult to slow it down once it is in motion. Its kinetic energy—the energy of motion—is proportional to its mass and to the square of its velocity.

But this simplified view of the situation should not mislead us into thinking that flying presents no problem to a bird. There are light, slowly flying birds and also heavy birds that must fly more rapidly. Heavy, fast-flying birds do have difficulties in landing. If their velocity is still high at the last moment, they plump down with a jarring thud. They are, of course, protected by the great prow of the breastbone and the massive flight musculature, but neither the skeleton and wings nor the skull, with the sense organs and brain, can withstand too strong a blow. During the courting season especially, when a bird is paying more attention to its intended mate than to any obstacles that may be in its path, smashing into a windowpane can bring injury or death—but not always. It is astonishing, but by no means uncommon, to see a small bird bang loudly against a window. tumble downward for only an instant, and then fly off in another direction. Birds that crash against the windshield of a moving automobile, though, are usually dead on the spot. A similar crash against an airplane windshield can do even more harm. Birds the size of a duck or larger can burst through the glass, causing a drop in cabin pressure and serious difficulties for the crew.

To protect themselves from the impact of landing, heavy, fast-flying, birds must choose a soft landing place. A large part of their momentum is absorbed in deformation of the landing place—as is particularly visible in the case of water or snow—and the rest by deformation of the bird's own body. The more the landing place is deformed, the more the body is spared. Soft landing places, however, are very rare in nature. One such yielding material is moss; I was thoroughly impressed by this fact one day when my brother fell out of a tree 5 meters high. He landed on his stomach on a thick cushion of moss and was not hurt. But moss cannot provide a useful landing place for rapidly-flying birds, since it usually grows in dense forests where these birds rarely fly. Moreover, it either disappears in winter, like most soft plants, or it is frozen as hard as stone.

There remains only water. Water presents no obstructions—it is uniform and relatively yielding, as we know from the landings of the astronauts or from our own experience in a swimming pool. Diving birds, in particular, have high wing loading and great difficulty in landing. In fact, there is an entire group of birds that can land only on water; on land they would do themselves too much injury. These are the large diving birds of the loon family, which includes the red-throated diver. This diver's approach to a landing is on a long, low path at very high speed. The feet are held far back along the body, so that the breast strikes the water first. With a braking force applied so suddenly from below, the bird would have to turn somersaults if it did not provide equal braking from above; it does this by tilting its extended wings to a steep angle. The diver does not always manage to strike an exact balance of forces right away, and a little instability is apparent—the bird rocks slightly back and forth. It slides about 5 to 10 meters and then comes to a halt.

Most water birds, such as geese, swans, ducks, and pelicans, land differently. They stretch their webbed feet forward and slide on them as though on water skis. But if there are waves or unpredictable winds it can happen that the landing becomes rather an inelegant sprawl. Now you may ask, "What harm does it do if the bird does fail to land as planned and turns head over heels?" It would certainly not be damaged, but there could be other disadvantages. Observe birds closely, and you will see that uncontrolled movements essentially never occur. Even in the most relaxed game, the most assiduous

On the following two pages: Diving birds like the red-throated diver (Gavia stellata, *lower left) make a "belly landing" on the water, whereas pelicans (upper left) land feet first. Both of them must slow down by raising the wings to match the braking effect of body or legs, in order not to tip over forward. When a mallard drake (right) approaches the water, it applies the brakes while still airborne, by raising its body upright, spreading out its tail, and vigorously beating its wings.*

preening—even in sleep—birds can be seen to be continually watching and listening to their surroundings. This helps to keep them from being surprised by predators. At the moment of a crash landing, a bird would be helpless.

So far we have considered the techniques used in the final part of a water landing. Usually a bird has no problem positioning itself for this final slide over the surface, but it can happen that the target is a small pond, surrounded by woods, which must be approached from a considerable altitude. How do the birds get to it? Take as an example a familiar sight, a landing mallard. These ducks often arrive at high altitude, and they try to reach the water surface as quickly as possible, since, during this rather difficult maneuver they are easy prey for swooping hunters like the peregrine falcon. Often ducks and geese can be seen to lay themselves on one side, in which position they lose altitude rapidly. But it is more common for them to descend slowly, with wings set at a large angle of attack. The ends of the wings are drawn slightly back, and the wings themselves bent somewhat. The bird, with the hind end raised a bit, glides diagonally down. One or two meters above the water surface it straightens itself up, and, with outspread feet, brakes its descent with the tail and by beating its wings, and begins its slide across the water. The way the wings are raised in this phase, and the large angle at which they are set, is shown in the photograph of two greylag geese landing. There are two reasons for raising the wings. One is that the area of the wing effective in generating an upward force is reduced, since this is proportional to the horizontal projection of the wing. In the case of the greylag geese, the reduction amounts to about 50%. As a result, the bird descends more rapidly. This may not matter very much, since at the time the wings are raised to this position the goose or duck has almost reached the water surface. The second reason is more important—the raised wings also provide stability during the slide over the water. The bird is less likely to tip over to the side. For example, as the bird begins to tilt to the left, the projection of the left wing onto the horizontal plane becomes greater. Suddenly the left wing is producing a greater upward force than the right, and as a result the bird automatically rights itself.

We have just been saying that lift is generated by the wings during a water landing, whereas before, the emphasis was on the drag that is produced—but the two are, of course, not mutually exclusive. It is simply a question of how much lift and how much drag. Certainly the proximal part of the wing, with its large angle of attack, generates chiefly drag and very little lift. The outer part of the wing, the manus, is warped forward so as to have a smaller angle of attack; there, more lift is produced.

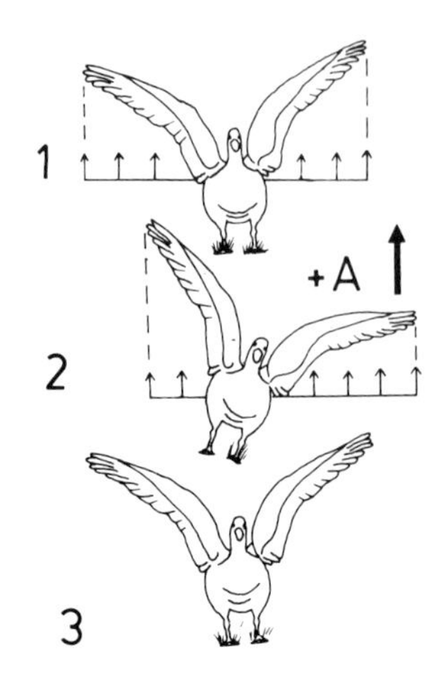

Greylag geese (Anser anser; *photo at upper right, sketch above) are stabilized during landing by the positive dihedral configuration of the wings. If the bird tilts to one side, the lower wing generates a greater upward force and the body rights itself. Thick arrow: upward movement of the wing. The number of thin arrows indicates the relative upward force on the two sides. In the lower photo a coot* (Fulica atra) *is landing. The coot is not able to slide across the water, since its toes are not joined by webbing (cf. p. 34).*

On the preceding two pages: Rosy pelicans landing on the water. The pelican in the large photo is having bad luck with the maneuver on this occasion.

Happy is the water bird with webbed feet! These birds are admirably suited for the long slide. The feet of coots are less practical for the purpose. The resistance necessary for rowing through the water in their case is produced by leaflike lobes on the toes. When the foot strikes the water surface, these are too flexible to permit a controlled glide. As a result, coots and gallinules often simply plop into the water, or they run a short distance over the water surface, decelerating at every step. At the same time, to avoid the danger of somersaulting, the birds beat their wings vigorously. This running-and-fluttering landing is not quite as elegant as the smooth glide of the web-footed birds.

When landing on water, a bird will never seriously hurt itself. Nor, I BELIEVE, HAS A DESCENDING BIRD EVER BEEN OBSERVED TO PULL UP INTO THE AIR AGAIN BECAUSE ITS LANDING MANEUVER FAILED.

Now we know how birds land on water. One thing is still lacking—a description of the actual process of braking in the air. An airplane is braked by lowering the flaps. This change in form of the wing is

A maneuver used by the sacred ibis (Threskiornis aethiopica) *to brake its flight. The ibis sets the wings at a large angle of attack. A similar effect is produced in the airplane above when the flaps are lowered.*

functionally equivalent to an increase in angle of attack. Birds increase their angle of attack in a different way; they tilt the leading edge of the wing up. If sudden strong braking is required, this can be dangerous. The wings can be set at such a large angle that eddies in the air on the upper surface of the wings ruffle the feathers. If this were to occur over the entire wing, the bird might go into an uncontrolled stall and crash. A "fine-control" system must be in operation to ensure a smooth descent—not to mention an accurate steering toward the landing point.

A mechanism patented by nature but copied by human aeronautical engineers takes care of this requirement. It is the alula, a group of feathers we have encountered before in the discussion of wing anatomy. Now we shall observe it in action. A goliath heron, braking its flight in approaching a landing place, has set its wings at so high an angle that the covert feathers on the upper surface are ruffled up. This is a clear sign that the airstream has separated from the wing surface at that place, that part of the wing is acting as a brake. In the outer, hand part the coverts lie close to the wing. There are two reasons for this. One is that, because of the twisting of the wing, the outer part is at a slightly smaller angle of attack than the inner. Second, even with a large angle of attack the raised alula causes the airstream to adhere to the wing. Since the air flowing over the wing comes mainly from the direction toward which the heron is flying, the angle of attack is approximately 50° in the proximal part and 40° in the manus. The air flowing through the gap between the alula and the wing is accelerated. (There is always acceleration when the space through which air flows is narrowed. Think how it feels to close a door against a draft; the air blows through more strongly when the door is almost closed than when it is wide open!)

When the bird is in the process of landing, the same effect prevents the airstream from separating even though the angle of attack is large. The alulae are always spread out when a critical angle of attack is reached, as can also happen during flight in a curved path. Braking is accomplished not only be raising the angle of attack, but by spreading the tail (and the webbed feet, in the case of swimming birds) and by tilting the entire body head-up.

Imagine that you were flying at an altitude of 500 meters, and that beneath you a tiny red spot appears, the roof of a house. You want to go there. How would you do it? Well, you could fly in a spiral, thus gliding down slowly, or you could float down more quickly with a parachute. But even the most skilled parachutist, under favorable conditions, would be unable to land at a specified spot on the roof. A stork does it every time. With wings set at a very large angle of attack,

Landing (above) and take-off (below) of a goliath heron. Although the positions of the wings with respect to the body are similar, the angles of attack differ; in the ascent the airstream comes from ahead of the bird and the angle at the manus is 20–30°, while in the downward glide (airstream from below) it is about 50°. In the upper picture air vortices over the wings raise the covert feathers; the raised alulae (visible on the right wing) act as subsidiary airfoils to help the airstream adhere to the manus.

the stork raises the alulae and sweeps downward. The wind rushes and roars past its wings, the proximal covert feathers flutter wildly about. But never mind—they are sturdy enough to take it. The bird hurtles toward the house and suddenly disappears. Has it crashed? It looks that way to anyone watching this performance. But then the stork comes into view again, shooting up from behind the house; it applies the brakes by beatinc its wings a few times with the manus warped during the back stroke, stretches its legs forward, and lands.

Why doesn't the bird land directly on the nest? Why does it first fly past the house, below the nest, and then up again? Birds of prey, swifts, pigeons, gulls, and many other birds behave similarly. When they want to land on an elevated place, they first fly a bit lower and make the final approach upward.

Certain birds do this even when they are going to land on the ground. For example, I have seen marabous flying very low over the ground, only to be carried up again on their outstretched legs during the actual landing.

Let us look more closely at this "climbing approach" to a landing, taking an auk as a particularly illustrative example. Landing points located at a height can be found everywhere. Most birds make their nests in the crowns of trees, in bushes, or on steep rocky cliffs, because such locations provide good protection from predators. The North Atlantic auks, diving birds with heavy wing loading, choose rock faces for their nests. The thick-billed murre is one of these birds. It approaches its nesting site, a narrow ledge of rock measuring only a square foot or so, on a steeply rising path. In the final moments, still rising sharply, the bird suddenly begins to beat its wings vigorously in a direction opposite to the flight direction. The wings, set at a large angle of attack, beat very rapidly and are pulled far forward. The hand part is warped during the upstroke in a manner typical of slow flight by large birds. The murre pulls its body upright, extends its legs toward the landing site, rises a little above it and finally drops onto the ledge from a height of 5–10 centimeters.

What is the advantage of such a roundabout landing? It must have something to do with the large wing loading and poor maneuverability of the murre, for its closest relative, the little auk, is a better flier and can also approach its target from above. The advantage is evident: imagine you were throwing a cup to your friend, once down from a third-floor window to the street and the next time from the street up to the window. When will he catch the cup more easily? Surely when you throw it upward, because then the cup is slowing down; as it rises gravity is constantly decelerating it. If it were perfectly thrown, its velocity would become zero at the very moment it comes within reach of your friend. But a cup thrown down is much harder to catch. It gains speed steadily as it falls; to the momentum imparted to it by the thrower is added the force of gravity, so that the kinetic energy of the cup constantly increases.

The bird's situation is just the same. A very maneuverable flier will be able to "catch" itself even if it is plummeting down to the landing place from above. Poor fliers, on the other hand, as are birds with large wing loading, could crash if they tried to do the same thing. This is why the murre lets itself be carried up to the landing place from below. Gravity reduces its momentum, and what remains is dissipated with a few well-judged braking wingbeats. This sounds very simple but definitely is not! Unpredictable gusts of wind can turn even a climbing approach into an adventure.

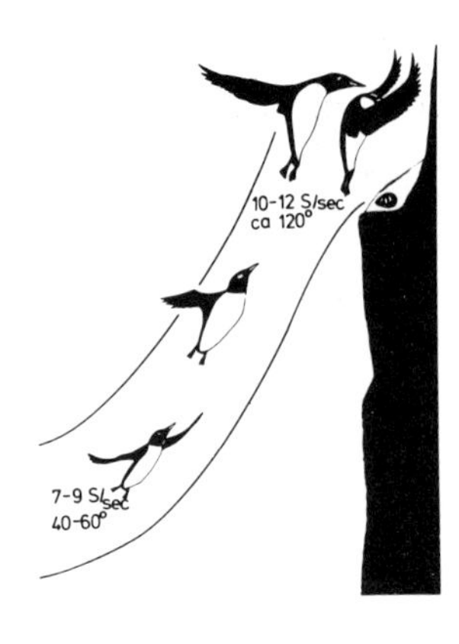

A thick-billed murre (Uria lomvia) *landing by "underflying." The wingbeat frequency and angle of the wingstroke are indicated in the sketch.*

Birds with large wing loading are not the only ones to encounter difficulty in landing on a confined space on a very steep rock wall; birds with relatively immovable wings also have problems. Among these are the fulmars, specialized oceanic soarers with wings that are long, narrow, very thick, and therefore rigid. All day long they soar over the ocean, fishing for food for their young. On Spitzbergen we were able to observe and photograph fulmars landing on the steep rocks where they breed. The proximal parts of fulmar wings are long and generate a great deal of lift during soaring. But the hands, so very important to a bird that flies by beating its wings and crucial for many maneuvers, are very short. Fulmars, too, use the underflying technique for landing. Just before they land, when it is necessary to use the wings to reduce the remaining momentum, they introduce a modification of the method: they turn the widespread wings on their long axes so that the underside moves quickly forward and back a number of times. This "wagging" of the wings, which amounts to a rhythmic raising and lowering of the angle of attack, provides an additional braking effect. Lift is generated as a by-product, and the birds can even steer themselves in a curve by wagging the wing on the inside of the curve through a greater angle than the outer wing. This makes the drag on the inner side higher, so the bird tends to turn around this inner wing.

While flying slowly up to its nest, the fulmar "wags" its wings, rhythmically enlarging the angle of attack (frames separated by about 1/25 sec).

Like the murre, the fulmar comes to a momentary halt in the air just above its landing place, with the body upright, feet outstretched, and wings set at a large angle of attack. Should the landing place be occupied or the maneuver fail, the fulmar must turn away and try again. It can also happen that a bird already on the landing place is driven away by the newcomer. The departing bird must try immediately to arrange its body so that each part produces the least possible

drag, with minimal protrusion of structures into the movement-generated airstream. The feet are instantly pulled back along the body, and the tail is pointed straight back. The wings are set at an angle that provides the most favorable airstream, and the bird soon reaches the speed necessary for good lift production by the wings.

In taking off, it is of course a great help to start from a high place. The bird can simply push itself out into the air and spread its wings in order to reach a speed that generates enough lift for flight. Taking off from the ground is a good deal harder. I have often seen murres making a great effort for minutes at a time without success. They whipped their wings against the water surface and made violent rowing motions with the legs until, finally, they rose into the air.

The feet and legs have a dual function during take-off. As accessory motors, they help propel the bird until a speed is reached at which the wings can take over. At the same time, they hold the body at a distance from the water, so that the wings can beat in the air. Flamingos, geese, and coots move their legs through the water as though they were walking, whereas pelicans shovel the water back with both feet at the same time. The wings help the legs to drive the bird forward, and they are the sole generators of lift toward the end of the take-off procedure. At that point the forward velocity of the bird has become so great that the legs, though they continue to swing back and forth, contribute a negligible amount to the motion; eventually they are laid back against

the body. The wings beat with powerful strokes, far forward and back again, so that the airstream across them flows as rapidly as possible. Whereas a murre, for example, beats its wings only 7–9 times per second when it is flying straight forward, during take-off it raises the wingbeat frequency to 10–12 beats per second. In addition, the wings are strongly twisted during take-off; the leading edges of the outer parts are twisted down, so that the wing section generates considerable thrust. The amount contributed to forward movement by the feet was demonstrated by a swan that was molting and had practically no primary flight feathers left on its wings. It could not fly and was being chased by two big boys in a boat. Its only recourse was to paddle with its feet, but by so doing it managed to shake off its pursuers.

Many birds have a hard time pushing themselves into the air from

A pelican can push off easily from a solid surface (left). When taking off from water, it must scoop the water back like a paddle streamer, simultaneously beating its wings far forward in order to generate adequate lift (below).

On the following two pages: Flamingoes taking off. Only after a rapid sprint and a succession of large wingstrokes have the birds produced enough lift to rise into the air.

the yielding surface of the water and prefer to take off from dry land. This was once illustrated by two pelicans; one was swimming in the water while the other sat over it in a tree. Whereas the first had to work strenuously for a time with wings and feet in order to take off, the other rose into the air in a moment.

Even on land, though, heavy birds and especially those with large wings must run for some distance before they can take off if there is no wind. Storks, geese, swans, albatrosses, and vultures cannot move their gigantic wings up and down fast enough to generate sufficient lift if they are standing still; air resistance and the inertia of the wings are too much for them.

At the cost of considerable exertion, I once obtained some insight into the fact that birds specialized for running, like the secretary bird, can take off only while running at high speed. Secretaries, which belong to the same order as the falcons and hawks, walk on their long, stiltlike legs through the steppe of eastern Africa, catching reptiles. A colleague of mine was determined that we should approach these snake hunters on foot, to take pictures of them. I agreed, but with

Birds with large wings like the secretary (Sagittarius serpentarius, *below) must create an airstream over the wings by running in order to fly. The ibis (right) can beat its smaller wings so far down that it can take off from a standstill.*

some scepticism. We crept quietly toward one of the birds and then began to run as fast as we could. But the secretary was far away in no time and disappeared behind the next hill. Later we followed it in a Land Rover. Then it had to hurry to get away, and at over 20 miles per hour it finally took to the air. Before taking off it ran for a while with wings spread out but not moving. Only at the very last, when it actually left the ground, did it begin to beat its wings.

These examples show that take-off and landing are relatively simple when a large "runway" is available. It is self-evident that starting from a rocky ledge or from a tree presents even less of a problem. Most birds of prey, such as buzzards, hawks, and kestrels, simply jump out of the nest with wings still folded and fall until they have reached the necessary speed for flight.

But it is much more difficult for birds that must fly steeply upward from a confined space. Helicopter pilots would be impressed if they could see some of the ways birds solve this problem. Even birds as heavy as pigeons and ducks can fly up vertically. They do so by using the same technique as in slow flight—thrust and lift are generated not only during the downstroke, but in the upstroke as well. The wings are

flexed and moved up or backward in such a way that the primary feathers are spread apart. Thus the airstream moves over the wing from behind. Each individual feather acts as a bearing surface, generating an upward and a propulsive force at the same time. This extra force production makes it possible for a bird with large wing loading to make a vertical take-off, land in the most restricted area, and even hover in one spot. Vertical take-off with warped wings is further assisted by the vigorous beating of the wings, so that lift does not decrease when the bird flies more slowly. Both the extent of the wing stroke and the number of strokes per unit time are increased. The sound made by the clapping together of pigeon wings when they take off shows how far the wings are raised and lowered at that time.

If it did not make use of these techniques, no pigeon, duck, or parakeet could take off vertically or descend straight down to a landing place. Because the problems associated with generating enough lift at a low speed are in principle the same in both cases, pictures of steeply rising and steeply descending birds look alike.

The smallest birds are exceptional in this regard. Small songbirds

Herons taking off, from right to left: the large-amplitude wingstrokes with manus inverted (second and third phases) provide sufficient acceleration for the heavy bird (montage of phases selected from different take-offs).

and hummingbirds are very light. Furthermore, they have relatively large wings. These two factors enable them to brake their flight rapidly. With only a few wingbeats in a direction opposing forward flight, they can come to a complete halt in the air or touch a landing place with their legs and thus absorb their remaining momentum. It is really astonishing how precisely small birds can land. The human eye cannot follow the event in detail. When a great tit flies to its nest box, it touches its feet to the lower edge of the entrance hole without stopping there. It has calculated its momentum exactly, and left just enough to carry the body right through the hole into the box.

Setting up drag surfaces, increasing the wingbeat frequency and the amplitude of the stroke, warping the manus during the upstroke, a climbing approach to a target, landing on water, dissipating momentum with the legs or with the well-upholstered body—these are the ways birds land. Falling away from an elevated starting point, propelling themselves forward with the legs, large-amplitude and rapid wingbeats, warping the manus during the upstroke, reducing drag by appropriate rotation of the wings and tail—these all help during take-off.

The heavier a flying object is, the more trouble it has during

The multiple-flash photo of a parakeet (Melopsittacus undulatus, *above) and the color photo of the parrot* Neophema pulchella *(left) show that when birds of this group are flying slowly they too use a very extensive downstroke and an upstroke with inverted manus.*

take-off and landing. Airplanes weigh more than 1000 times as much as birds. In order not to crash, they must get through the slow parts of the braking or acceleration processes on the runway. Birds have accessory methods of staying in the air even when flying slowly. But even birds differ in their abilities to take off and land. The possible reasons for this variability will be considered in the chapter "Bird Flight as an Adaptation to the Environment."

Up and Down in a Matter of Seconds

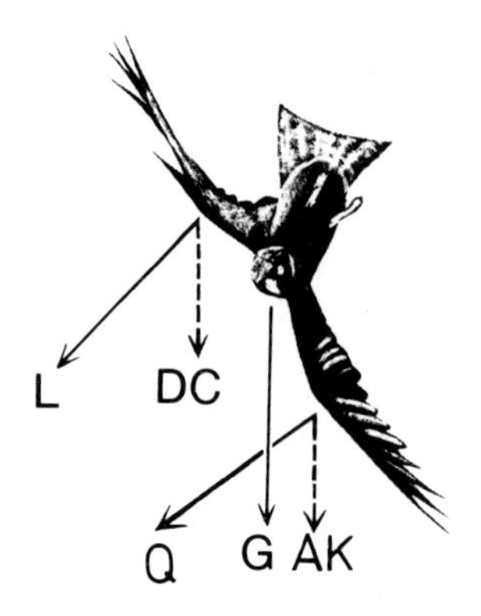

A black kite swooping downward on its back; because the wing is now convex down, lift exerts a downward pull, as shown in the sketch: L, lift; DC, downward component of resultant force; G, gravity. The photos show the kite swooping (upper left), soaring (upper right), preparing to seize prey (lower left), and scanning the field for prey (lower right).

Birds know nothing of the laws of physics, but they follow them faithfully as they move through the air. To gain altitude rapidly, it is necessary to be light in weight and have strong flight musculature. Small birds that meet these requirements are therefore the true masters of vertical upward flight. In theory, the fable of the eagle, which flew the highest, and the wren which rode along secretly and then flew even higher, could actually have happened. Small birds can fly very steeply up but cannot fly for very long. The power of their muscles is relatively large, but they use up more energy and soon tire. Large birds, on the other hand, can reach greater altitudes because they have more endurance and, even more importantly, because many of them can soar.

In the vertical take-off, though, small birds are unbeatable. No large bird can move straight upward so rapidly, over a short distance, as a small bird. A pigeon or duck can also rise almost vertically for a few meters, but it is a laborious procedure; sparrows, by contrast, can fly straight up to a roof 10 or 20 meters high with no trouble at all. They in turn are excelled by larks, which can rise steeply to altitudes of over 200 meters. Larks make particularly long upward flights when there is a slight headwind. They never furl their wings completely, but rather, let the proximal parts act as lift-generating surfaces even during the upstroke. This generates a force pushing the bird backward, as also occurs during fast flight by large birds. This backward force is compensated by an equally large propulsive force during the downstroke.

Like many other small birds, the lark can rise more rapidly than large birds. But the larger birds are superior in steep downward flight. It does look quite fast when a lark folds its wings and falls like a rock. Its kinetic energy is very small, so that it has no difficulty in stopping its fall just above the ground by simply spreading its wings. But large birds, too, fold their wings when they want to descend rapidly, and because of their greater weight/drag ratio, the free fall is a good deal faster. Peregrine falcons are said to swoop down at speeds of over 200 miles per hour. At such times the outline of the body is a rounded triangle and its longitudinal section is an ideal drop shape.

I observed another technique for high-speed descent when watching black kites in the Ngorongoro Crater in Tanzania. At one of the places set aside for tourists, a group of travelers thought they could have a picnic in peace, out of reach of the distant lions. But they did not reckon with the danger from the air; a kite swooped into the middle of the horrified crowd and seized a piece of chicken. Bloody

A LITTLE TOO EXTREME!

scratches on the back of one traveler's hand bore witness to the accuracy of the kite's aim.

The kites usually circle in the air, waiting, or perch on trees so as to take immediate advantage of any chance of food. In order to get pictures of them, I offered a piece of meat to two kites circling above me. Both birds saw me, and a real race developed between them. While one of the kites plunged down with wings retracted, the other threw itself onto its back and shot downward even more rapidly. Just

Above: A herring gull beginning its swoop by tipping its body forward with the legs splayed and the wings raised high.
Right: A laughing gull pulling out of a swoop by fanning its tail and flicking it up.

above the ground it rolled once again and seized the bit of meat; it won the race by a wide margin. While its rival was depending entirely on the effect of gravity, this kite was able to accelerate its fall initially by the lift generated by its own wings. After the 180° rotation, the convex wing surfaces pointed down, and therefore so did the lift due to the airstream from in front of the bird. The movement was so fast that it was almost invisible to the eye, but with a great deal of luck and an exposure time of a thousandth of a second I was able to capture this rare maneuver on film. This form of locomotion was certainly not a

trick known only to this one kite. Other kites, and even birds of other species, have been seen to do it.

Geese and ducks, for example, shorten the often dangerous phase of downward gliding from great heights by a maneuver that looks equally risky. They throw themselves on their sides, or even all the way onto their backs, and like the kites, swoop down with the greatest speed. A pursuer would have a hard time catching a bird plunging at this rate.

Airplanes usually crash after colliding (above), but birds do not, as illustrated by the photos of vultures (upper right), eagles (lower right), and gulls (cover).

Flight is Not Only for Locomotion

Collision in mid-air—and crash! This is the nightmare of every pilot. When human aircraft touch one another in the sky, the consequences are almost always serious. Now that the airplane has become one of the most important forms of transportation, we often hear of such catastrophes. And near misses have become almost a daily routine.

Flying birds almost never bump into one another unintentionally. Birds can change course, slow down, or accelerate in the very last moment. Intentional touching or bumping, on the other hand, are an integral part of aerial courtship or aggression in flight. If one bird wants to drive another away, it simply flies up to it and relies on the sudden rapid approach to frighten the other and make it retreat. This is usually what in fact happens. But occasionally the threatened bird stays stubbornly where it is, so that the attacker has to swerve away at the last moment. This threat of collision is made particularly often by birds that need to secure a good perch—gulls waiting for a chance at a mast, an osprey or sea eagle looking for a tree branch overhanging the water, buzzards seeking a pole, or male storks competing for a nest.

It is interesting that birds like auks, pheasants, and ducks, which are not skillful enough to avoid collision in the last second, do not use this method of threatening rivals. The point of the maneuver, after all, is not to do real damage but simply to make a frightening impression. Birds fly more seriously to the attack when they are fighting for a territory or a female—and, of course, when hunting prey. Another form of threat in flight is called mobbing. This is a group activity in which songbirds turn on their enemies—birds of prey, cuckoos, jays, and, most commonly, owls. The fling themselves at the foe with loud screeches, but usually do not go so far as to injure the target of their attack; rather, they simply try to put it to flight with a shrill uproar and wild aerial acrobatics.

By swooping vertically down, terns, arctic skuas, and herring gulls, as well as certain owls such as the tawny owl, can drive away even quite large enemies with their pointed beaks and knife-sharp claws.

On the following two pages: The combined swoop and dive of the Arctic tern. It plunges down head first and catches its prey just below the water surface. The montage shows different sequential phases in this maneuver, from upper left to upper right.

This form of defense is especially well developed in species that breed on the open ground or live in other poorly protected places.

For a bird to threaten successfully, drive off, and even injure another bird while flying, its flight technique must be superior in some respect; it must be faster or more agile or have weapons better suited to aerial combat. An arctic skua is more maneuverable than the larger glaucous gull. As a result, it is able to repulse this egg thief. But when two animals of the same species fight, their flight capabilities are basically the same. In this case the outcome depends upon the individual's skill in performing the various flight maneuvers or proficiency in the use of the weapons available to both. If neither has a clear advantage, battles aloft can be fierce indeed. One can often observe colony-breeding birds like terns and auks locked in a grim struggle. Each in the clutches of the other, a pair of auks will tumble down until they splash into the water.

In Uganda I once saw two hooded vultures fighting above a feeding place. One was trying to dig its sharp claws into the bare, unprotected head of the other. The animals fought with necks stretched far out to the side so as to escape the opponent's savage weapons. Why were these vultures dueling in the air? Surely they could have done as well on the ground. No, they are better off where they are—in the air escape is easier and any blow that does land is likely to do less damage.

By contrast, when birds like ospreys, sea eagles, falcons, and hawks swoop down upon their prey, the impact must be as great as possible. The more harm done to the victim, the more energy the predator saves and the sooner it can begin its meal.

It was hailed as a great achievement in aeronautics when a way to refuel jet planes in mid-air was developed. But birds have been doing the same thing for millions of years. I am not referring to the swift, which simply opens its beak and lets insects fly into its outsize maw, but rather to birds of prey. Peregrine falcons and kites throw prey to one another in flight, particularly during the breeding period. The recipient rolls onto its back just long enough to catch the morsel in its talons and then lands in order to eat it. Once, while watching marsh harriers, I thought at first that a battle was in progress when I saw two distant birds performing similar maneuvers without any prey. When they came closer I could make out the details. A male was swooping down toward a female, which threw herself onto her back just before they met and stretched her claws out as though in self-defense. But there was no real aggressive intent on either side; the whole display was part of a courtship ceremony. Sham battles involving maneuvers like rolling onto the back and attack by the male are performed by various birds of prey—the bearded vulture, the black kite, and other

The frigate bird Fregata magnificens *is the pirate of tropical seas; here some of them are swooping down upon a red-footed booby* (Sula sula) *to rob it of its prey.*

From left to right: a greylag goose is being chased by a rival which it cannot easily shake off on the water. But on the ice (far right) the quarry escapes by making a sudden turn that the rival cannot follow so

kites. In the process the birds can hook themselves firmly together and spiral awkwardly toward the ground. Barely in time to avoid a crash, they break apart and pull out of their fall.

A single flying bird can also produce impressive courtship displays and territorial rituals. I once watched a male marsh harrier spiraling up to altitudes of several hundred meters, where it gave a real aerobatic performance. With wings swept back it shot here and there, looped the loop, swung to left and right—all this with sudden intervals of quiet soaring or earthward plunges—until, after several minutes, it flew back down. During the return flight it alternated between falling with furled wings and tumbling along a spiral path. At last, swaying from one side to the other with wings half flexed, the bird fell into the reeds and rejoined its partner. Kites indulge in similar courtship displays.

Hawks and kestrels "weave a garland" in flight, making upward and downward movements in alternation. Booted eagles fly to great altitudes, shoot down as much as 800 meters, pull out and rise again to begin the game anew. Whenever a bird does not appear to be hurrying straight toward some goal, but rather flies conspicuously about, one can justifiably regard this activity as a courtship flight. By "conspicuous" flying is meant anything that departs from the norm. There are changes in the flight path and in the way the birds move, as we saw in the above examples. Snipes fly along an arched path, during the descent vibrating the fanned-out tail feathers so as to produce a bleating sound. (They have been called "sky goats" because of this behavior.) Redshanks fly over less extensive arcs, rushing upward with rapid wingbeats and loud calls and then gliding back down with wings outstretched to a tremolo accompaniment of clear notes. Tree pipits behave similarly; they float singing down from the sky almost as though suspended from a parachute. Hummingbirds descend like

quickly, despite full braking with wings tilted to a great angle and tail dragging behind.

bullets, whizzing 20 to 40 meters downward at a speed of over 60 miles per hour. Pigeons flying up with wings beating at a high rate and then gliding down are probably a familiar sight to everyone.

Another way birds make their flight conspicuous is by a great departure from the normal wingbeat frequency. Tits beat their wings considerably above the normal rate and at the same time reduce the amplitude of the stroke, so that a buzzing sound is produced. Unusually slow wingbeats are used by greenfinches, linnets, and black-tailed godwits; with wings outstretched and beating languidly they roll from side to side. The lapwing, which during migration flies in almost a straight line, becomes a whirling dervish in its breeding territory. Sweeping its wings through a large angle, it turns precipitously; sudden falls through the air alternate with lunges straight upward. The wings make a sound as though a rod were lashing irregularly about. The African widow-birds, songbirds with long tails, disdain horizontal flight when they are courting. They dance up and down across the steppe like yo-yos.

Courtship flights are both beautiful and intriguing. But the main point of interest here is that they reveal so impressively what birds can achieve in flight. Airplane pilots on scheduled routes are normally compelled to fly economically, in straight lines. But at an air exhibition they show off everything they can do. Birds are just the same, in that they normally hasten timidly through the air. When they display sudden plunges, vertical ascents, loops, rolls, tumbling here and there, or maneuvers made risky by their very slowness, they too are trying to make an impression—on their females. But sometimes one can imagine that they behave this way simply for the fun of the thing. For example, on particularly favorable days, when there are updrafts or headwinds, a flock of gulls or crows will rise into the air and then, with no apparent reason, interrupt their calm flapping or

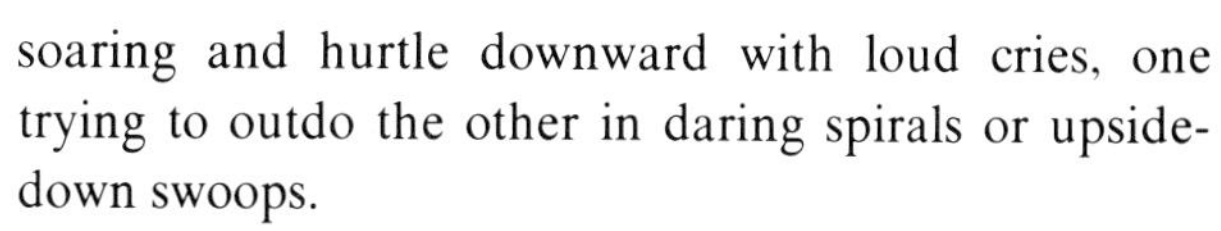

soaring and hurtle downward with loud cries, one trying to outdo the other in daring spirals or upside-down swoops.

For a jet airliner to travel safely, the constant alertness of several crew members is required. By contrast, the thought of flying beings that sleep aloft is alarming. But swifts have been said to be capable of flying in their sleep. Of course, no one has ever observed this directly! The evidence amounts to their having been seen flying at great altitudes in the evening and doing the same thing the next morning.

Some elements in the courtship flight of marsh harriers (Circus aeruginosus)*: sham battles (far left; the upper photo shows the female on her back), swooping at a great height (top photo at left), whirling downward (middle), and weaving back and forth (bottom).*

A Million Hours of Flight Time Without a Crash

An army of technicians, ground-control specialists, scientists, laborers, administrative staff, and, not least, the crew keeps watch over the safety of airline passengers. Teamwork evidently pays off, for small private planes crash more often. When a complicated flying machine m26is under the control of only one person, any mistake he may make has immediate and dangerous consequences. Birds fly with no outside assistance, but in spite of this they hardly ever crash. I have certainly observed more than a million flying birds and have seen only a few come to grief.

The first time it was my fault. I had joined some other children in taking magpies from the nest (please don't follow our example!) so that we could compete in raising them to adulthood. Which bird had the finest appearance in the end, and which flew best? None of them! We 12-year-olds of course, had no idea which food would be suitable for these animals. Things like vanilla pudding from our mothers' kitchens were poured down the poor creatures' throats. We wound up with fat magpies quite unable to fly; their wing loading was so great that they fell to earth like stones.

The second bird I saw crash was a young glaucous gull. It was making its first attempts to fly over a rocky pinnacle high above the Norwegian coal-mining town Longyearbyen on Spitzbergen. A sudden gust of wind unfortunately struck the bird in such a way that it could not keep its balance. It fell a couple of feet down to its nesting place, a ledge of rock only a square foot or so in area.

Finally, I have watched murres, with their high wing loading, crash on several occasions. On Spitzbergen a murre was trying to land on a pointed rock. It misjudged its approach, failed to apply the brakes in just the right way, flew on past the projection and slipped through a crack in the rocks. Ten meters below it suddenly appeared again, and 20 meters further down it hit the slippery rock face. On the island of Helgoland, too, I once saw a murre crash into rock from an altitude of about 15 meters. In neither case did the birds seem to have suffered much from their fall, for they flew out over the open sea immediately afterward. Birds with high wing loading can crash from time to time. But what is that, compared to the millions of hours of perfect flight!

As a general rule, even birds with high wing loading have reserves of strength that permit them to beat their wings faster in a difficult situation. Another safety factor that affects most birds' ability to fly is the size of the wings. Even when some feathers are missing, most birds can still fly quite well. Konrad Lorenz cut feathers off pigeon wings, to see how many had to be lost before the pigeons became unable to fly.

The results were astonishing. With the primaries almost entirely removed they could still fly well horizontally but could no longer fly upward. When the secondary feathers were much shortened, on the other hand, they could ascend but not fly horizontally. I tried a similar experiment with pheasants, removing the first five of the ten primaries. In this condition they could actually develop a higher starting velocity than a hawk taking off at the same time.

CHAPTER SIX

Bird Flight as an Adaptation to the Environment

A Proposal for the Design of a Bird

A fulmar cannot hover or fly in the woods. A small, round-winged songbird, on the other hand, is very agile and can quickly slow down and accelerate—but it hasn't much endurance. A heavy bird has problems in landing and making tight curves. A pheasant takes off like a rocket but then tires out and soon must land again. In a hurricane over the ocean a stork is whirled helplessly through the air . . . What a list of inadequacies! I want to try to construct a better bird, a universal type of flier that can do everything.

Perhaps the limitations on flight in the different bird species are related to the size of the birds?

To find this out, we shall first build a very small flying animal. Let's make one only half as large as the smallest existing bird, the vervian hummingbird, which has wings only 2 centimeters long. Since small flying animals must beat their wings rapidly, we choose a wingbeat frequency of 300 beats per second. But with the wings beating at that rate, neither the driving power of the muscles nor the control by the nerves can keep up. Each muscle fibril and each nerve fiber needs a phase of recovery and replenishment after each phase of activity, and such a rapid sequence of wingstrokes simply does not allow a pause.

It looks as though our plan of building a small bird has been frustrated. But perhaps we can still succeed by using an elastic type of construction, so that the kinetic energy of the beating wings is not lost but rather, remains available for the next wingbeat. A solid rubber ball can bounce up several times for each time it is thrown to the ground.

In fact, such a driving system has been evolved by the smallest flying animals, the insects. Many insects fly by means of a click mechanism that can be illustrated as follows. Take a piece of stiff material (tin, for example) held under tension in such a way that it arches slightly upward. Sufficient pressure on the upper surface will cause the metal to spring down with a loud snap; now it arches downward. If the kinetic energy of the tin, as it jumps from one position to the other, is returned to it by an elastic material, it can flip back and forth several times.

The bodies of insects consist of stiff walls. The front part of the body is roofed over by a plate connected to the two side walls by an elastic joint. The covering plate actually does swing up and down several times once a single pull by the muscles has set it in motion, and the wings simply oscillate along with it. The wings are attached to the plate in such a way that they behave like two spoons stuck under the too-small lid of a cooking pot. If the lid is pressed down, the spoons (wings) move up, and when the lid is raised they fall back down again. In an insect there are muscles which produce these effects. One set of muscles is arranged longitudinally from front to back in each segment of the thorax, the part of the body that bears the wings. When they contract, because of the position of the strands of muscle in the upper third of the body, the upper plate (the "lid") arches upward, and a downstroke of the wings is produced. The antagonists of these longitudinal muscles are two groups of muscles running from the upper plate to the underside of the body. These pull the arched plate down again, as a result of which the degree of arching is reduced and the wings beat upward.

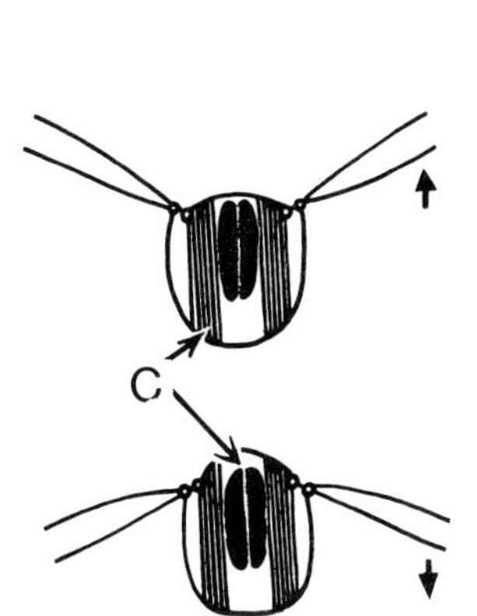

Diagram of the wing "motor" in insects with indirect flight musculature. Contraction (C) of the dorsoventral muscles (vertical lines) pulls down the plate covering the thorax, and the wings swing upward (arrow). When the longitudinal muscles (black) contract, the dorsal plate arches upward and the wings are lowered.

Insects are the only animals that have used this click mechanism to produce high-frequency oscillations of body and wings. They have exoskeletons suitable for the purpose. But it is impossible for a bird, with its internal skeleton, to beat the wings faster than the muscles and nerves can drive them. In our effort to construct a bird that meets as many of the requirements for perfect flight as possible, therefore, we can rule out small models. Perhaps very large birds are the sort of thing we are looking for?

We can immediately see that one important property is lacking; large birds are not very maneuverable. Sudden changes of direction within a very confined space are not within their repertoire. If maneuverability were to be increased by the addition of extra braking and propulsive apparatus, more muscles would have to be installed. But these would increase the total weight so much that the wings would have to be enlarged in order to support it. To move these larger wings, the muscle mass would have to be still further increased. There is the same problem with the wings that we previously found with bones; the rapidly increasing weight of a growing body can no longer be supported. As body size increases, the area of the wings increases much more slowly (as the second power) than does the weight of the body (as the third power). The largest birds capable of flight have evidently already reached the greatest body weight that can be borne by muscle-powered wings. So, in constructing our new type of bird, we cannot go beyond the existing range of sizes.

Perhaps we shall have more success if we combine the flight characteristics in which each of a number of birds excels! We'll give the fulmar the ability to hang in one spot from its vibrating wings, like a hummingbird. To do this, it would have to have shorter, more movable wings—but with these it could no longer glide, scanning the water surface all day long for food. This proposal, then, is also impracticable.

On the other hand, if small round-winged birds are to be converted into more rapid fliers, they must become heavier so that their greater kinetic energy can help to compensate for drag. But this would be possible only at the price of a reduction in their high maneuverability.

Another way to fly at high speed in spite of being small has been developed, for example, by the barn swallow. These birds have rather long, pointed wings. Long wings produce a greater propulsive force than short wings, since they travel over a longer path at each stroke and therefore generate a stronger airstream. Barn swallows are among the best high-speed fliers. However, they are incapable of thermal or dynamic soaring, and in dense shrubbery they are far surpassed in agility as well as in endurance by, for example, a flycatcher.

If we wanted to turn the stork into a bird that could also fly over the ocean, its wings would have to become narrower and thicker (more wind-resistant). Then, however, it would have lost the capacity for thermal soaring and would be unable to reach its winter quarters. Chickens and pheasants could be "remodeled" to permit long-distance flight only at the price of their sprinting ability. Then they would be easy prey for hawks.

This approach to the problem—the combination of several skills in one model bird—has also failed. As we have seen, the characteristics of different kinds of flight cannot be arbitrarily exchanged. The reason, of course, is that each is adapted to the particular conditions in various habitats.

Flamingoes searching for food (right page) and landing (below).

How Flight is Adapted for Feeding and Repelling Predators

Apart from the wind, the effects of which we considered in discussing soaring over land and sea, two other factors have certainly affected the development of the ability to fly—the nature of a bird's fkod supply and the necessity of protecting itself from predators.

The way each kind of bird has adapted to these aspects of its environment is reflected in the way it flies. Let us begin with the birds of eastern African lakes. Flamingoes are particularly common there. These long-legged birds, relatives of the storks and herons, seek their food in the shallow water along the shore. With their beaks they strain algae from the surface water. Thanks to their long necks they can reach the surface even when the water level is low. They breed on earthen hills that they build in the shallow water where it is hard for land predators, most of which are reluctant to enter water, to reach them. The breeding site, then, is relatively safe. Only occasionally does the Cape otter, a predator adapted to aquatic life, succeed in attacking and killing a flamingo.

I once had an opportunity to watch an otter raid a flock of flamingoes. Soundlessly and almost invisibly, the otter had crept up to the pink and white crowd. Suddenly it sprang into the middle of the group of birds and pulled down a flamingo, fluttering in terror. Gripping the bird's bill firmly in its teeth, it dragged the bird toward the shore despite the furious flapping of the wings. The flamingo was transported in this way over more than 100 meters. It didn't have a chance. Growing ever weaker, it gradually ceased to struggle. The

The Cape otter has caught a flamingo.

otter held the head fast and repeatedly tried to push it under water. Finally, the otter hauled its victim on land, up a steep bank, and into the underbrush. But it happens all too often that an otter fails to steal close enough without being noticed, for it must escape detection by several thousand pairs of watchful eyes.

Another advantage to the flamingoes of a breeding site protected by shallow water is that it is easy to reach. There are long unobstructed runways for takeoff and landing. Any flamingoes standing in the way will move to the side when another flies down to land. When the food source is directly in front of the door, and when relatively safe breeding sites are easy to reach, then there is no need for the birds to be particularly good at flying. And in fact flamingoes, with their

relatively high wing loading, fly quite awkwardly.

There are other long-necked stilt-legged birds that also feed in the shallow shore zone—storks and herons. We can recognize them at some distance by their more serene appearance in flight. Their wing loading is less than that of the flamingoes, and they can take off without running and land precisely, down to the last centimeter, even on moving branches. Because of their skillful flight they can raise their young in trees or in hiding places among reeds, where the small birds are protected from foxes, civet cats, snakes, and the like. A flamingo trying to land in a tree would inevitably crash. It cannot fly slowly enough.

Another example shows how important the environment can be in determining a bird's ability to fly. The cormorants on the Galapagos Islands have no predators. Therefore, they are never forced to flee, and they can choose breeding sites in quite unprotected places. Since the ability to fly has become superfluous, they have lost it. It could be that this development will be as fateful for the cormorant as it was for the great auk, which also became unable to fly. As a result, that auk could no longer cope with the new environmental factor "man," and it was because of man that the giant bird became extinct.

The need to find food has also had a crucial effect on the evolution of flight in many species. For example, birds that bring their food up from deep underwater must either be heavy or be able to plunge in a rapid dive from the air into the water. Both lines of evolution have been followed, and as a result, completely different ways of flying have been developed. Moreover, for each of the two types of diving, several different modifications of flight can be found. As a rule of thumb—with only a few exceptions—it can be said that the better a bird can dive, the worse it flies.

In the absence of land predators, the wings of the Galapagos cormorant (Nannopterum harrisi) *have regressed; the bird has lost the ability to fly.*

When night falls in the Serengeti, the marabous retire to special trees to sleep.

As a first example, consider the auks—heavy birds that dive well. They live in the northern oceans, where food is abundant. There, vertical currents carry nutrients up to the well-lighted surface waters, where green plants use them to synthesize organic matter. Thus begins a food chain of enormous productivity, extending from the plant and animal plankton to the fishes. This wealth of food is strictly divided among the four auk species living there. The thick-billed murres hunt the large fish, which swim the fastest. Puffins catch smaller fish, black guillemots search out bottom-dwelling animals near the shore, and the little auks or "dovekies," no bigger than starlings, eat slowly swimming crustaceans or pelagic molluscs. To capture their prey, the murres must be the best divers; dovekies are slower and not so adroit

when swimming underwater. The latter can afford to take more time catching their prey, since it moves more slowly. The difference in diving ability among the auks has had crucial consequences with respect to flight, for all auks move underwater by means of their wings. Partially flexed, the wings are moved back and forth like oars. To overcome the resistance of the water, auk wings are small and relatively thick. This shape is an advantage for rowing during a dive, but is not very suitable for flying. Murres—the best divers and simultaneously the worst fliers—have the greatest wing loading, 260 grams per square decimeter; the puffin has a wing loading of 220 g/dm^2, the guillemot 180 g/dm^2, and the dovekie 120 g/dm^2 (from the

Murres, with their heavy bodies and small wings, are only moderately good fliers, but strong and agile divers. (Upper left, murres on their breeding rock; upper right, flying away.)

The dovekie, a close relative of the murre, is a poorer diver but better at flying. (Lower left, dovekies taking off from a rock; lower right, landing.)

The wing loading of the red-throated diver is large, so that it can land only on water. It breeds on lakes and ponds, nesting on little islands it builds of moss (right). The photo above shows the observation tent of the author and a brooding diver on Spitzbergen.

data of Kartaschev and my own estimates). What these wing loadings suggest, observations confirm.

Murres have great difficulty taking off from the water when the air is still or there are no waves (to serve as starting ramps). For this reason they usually prefer to escape, when necessary, by diving. And when flying up to a protected rock face they make more unsuccessful attempts at landing than the other species. Puffins, on the other hand, can even drop into their holes in the rocks from above. Black guillemots fly with no trouble to the top story of the breeding cliff, where they nest. But the best fliers are the dovekies. They make completely effortless landings in the most difficult places, braking their flight with only a few wingbeats. They can let themselves gently down onto a landing point with no need of a climbing approach. Since they fly so well they can easily reach breeding sites located far inland. Dovekies are also the most common species of bird in Spitzbergen.

The extent to which specialization for underwater capture of prey can be taken is demonstrated by another bird of the same habitat, the red-throated diver. It uses different means of propulsion while hunting. With an effect similar to that of a ship's screws, its feet, attached to the body at its extreme hind end, drive the body forward. In this way it can cover several hundred meters underwater in a few minutes. As effective as the feet of this loon are underwater, on land they can barely support the heavy body—and certainly not during a landing. Thus the safe breeding sites on a bird rock near the sea are inaccessible to a red-throated diver. Instead, it builds islands of moss on fresh-water lakes or the banks of rivers, where it can land on water. But, since such places are quite rare, the distribution of this bird is very limited.

Penguins can dive more expertly than auks and loons. Their wings have continued the tendency toward size reduction, as exhibited by

the auks, to such an extent that they have become entirely converted to oars. But penguins have relinquished the ability to fly, and have become swimming and diving birds, for reasons other than the need to obtain enough food. The predation factor was also at work. For one thing, there are no land predators in the Antarctic, so flight has not been necessary. Secondly, the Antarctic, has many aquatic predators—the leopard seal and killer whale—which the penguins escape by divihg.

In the polar waters where these birds live there is so much food that a bird need only swim to find it. By contrast, birds in regions where food is scarcer, as in the tropics, must cover much greater distances. If the birds making these prolonged flights are to be able to spot their prey while still aloft, they must be masters of slow flying. A prerequisite for slow flight is low wing loading. But for light birds to be able to go under the water surface far enough to catch their prey, they must enter the water with high momentum. That is why these birds have developed the second method of underwater hunting, the swooping dive. Of all the species that begin their dive for prey while still in the air, the boobies are the best divers. They are the only birds that can catch fish at considerable depths by this technique. Brown

Above: Leach's petrel (Oceanodroma) *feeding at the water surface. With graceful motions &the bird flies in hops over the water; its legs, which cannot support it for long on land, keep it the right distance from the surface, so that its wings do not get wet.*
Right: The penguins (here, the rock-hopper penguin Eudyptes cristatus) *have practically become aquatic animals; their wings, no longer suitable for flight, are used as oars.*

pelicans, terns, kingfishers, ospreys, and sea eagles keep watch for fish swimming close under the surface and usually hunt only in shallow water near the shore.

In order to dive so deep, the booby must accelerate greatly as it swoops down (I have measured speeds of up to 110 kilometer/hour). Let us watch a hunting bird at work. Blue-footed boobies swished arrowlike around our small boat on the lake. We had drifted into the middle of a swarm of fishing boobies. After each swoop into the water, the birds bobbed up again like corks. Then they took off, running briefly along with their feet just on the water surface. Following an arc with radius about 50 meters, they rose to an altitude of 30 meters to develop momentum for the next dive. In shallow bays we had noticed that they plunged in from much lower altitudes, 2.5 to 10 meters. Whenever a booby slowed its flight, we knew that it was about to dive. To get into swooping position the booby either turned onto its back, simultaneously snapping its tail back, or tilted one wing down and slid off to that side. In the latter case, the downward wing was more strongly flexed than the other and set at an angle such that the airstream struck its upper surface (a negative angle of attack). While hurtling downward the birds accelerated still more by beating the wings a few times. Then they pulled the outer parts of the wings further and further back. In so doing, some of the birds rotated several times about their own axes. Just above the water the boobies whipped

their wings back so far that, like slender, streamlined rockets, they broke the surface almost without a splash. The dives carried them deep into the water; dives to as much as 20 meters have been observed.

It is not quite clear how they catch the fish. The suggestion has been made that they approach the prey from below, on their way to the surface, steering through the water with the wings, legs, and tail. I regard that as unlikely; as they strike the water and glide down through it, the disturbance would surely drire the fish away. But it seems clear that the birds do not go a very long way horizontally underwater, for they always come up again only a few meters from the point of entry. The many swoops through the air that are broken off at the last moment seem to me to indicate that blue-footed boobies take aim at their prey while they are still in the air. And after all, the great advantage of this method of hunting is that it parmits a surprise attack. When pounced upon in this way, even rapidly swimming fish can be captured, though they would surely escape a bird swimming

Blue-footed boobies (Sula nebouxii) *diving from the air into a school of fish.*

after them. Their specialization as "torpedoes" has become the basis of the boobies' existence.

Another exception to the rule "good diver = poor flier" is the anhinga. This bird, closely allied to the cormorants, can fly well and dive superbly. Its feathers let the water through, so that when the bird swims, all the air is driven out and it sinks to considerable depths. Swimming anhingas look like snakes; only the head and neck show above the water. Under the surface they row with wings half extended, and using the long neck like a flexible S-shaped spear with the beak as its tip, they stab the fish in their path. Like cormorants, anhingas dry their feathers in the sun, spreading their wings. When its feathers are dry, the anhinga is an astonishingly good flier. At Lake Nakuru in eastern Africa they can be seen soaring to great heights along with the pelicans. No other diving bird can do this. The wetting and drying of the feathers make such a transition possible.

The need to secure an adequate food supply has been particularly effective in bringing about special adaptations of flight among the birds of prey. Because these birds specialize in certain animals as prey, each species has had to develop appropriate hunting methods, and these differ from one another primarily with respect to the type of flight employed.

But less specialized birds of prey also exist. For example, owing to their size eagles have not been able to develop a special, highly effective hunting technique. Their large wings cannot be moved very rapidly, and the big body does not permit quick braking, acceleration, or turning. A sea eagle, therefore, uses various methods. Often it hunts while sitting still, like a buzzard, scanning its surroundings from an elevated look-out point and swooping down from there when a likely victim appears. On the other hand, a sea eagle can frequently be seen on reconnaissance flights, during which it can stay suspended for several seconds over one spot, peering down at the water or ground. With a brief vertical flight it can even hurl itself upon standing herons, storks, or cranes and carry them off as prey. Eagles have actually been seen to hurtle down for altitudes of several hundred meters and submerge completely in the water. By doing this they can capture very heavy fish. Fish weighing 12 to 15 kilograms, too heavy for an eagle to fly away with, it brings to shore by swimming. The sea eagle forces coots to dive with attacks repeated so often that eventually the coot, exhausted, gives up and can easily be seized on the water surface. When eagles are hunting in this way, they may actually work together in pairs. While one sweeps low over the water, compelling the coots it finds to dive, the other hovers on high, only to strike like lightning as the coot comes up again.

On the two following pages: Because its body is so well streamlined on entry into the water, the blue-footed booby dives deeply. The montage, from top left to top right, shows the dive through the air and into the water, followed by emergence and take-off.

Anhingas (Anhinga rufa) *are excellent divers and good fliers. Since their feathers are water-permeable, after a swim (bottom) they must dry out (middle) before the bird can rise into the air again (top).*

Sea eagles also behave as "food parasites," stealing fish from the gulls that caught them. Moreover, they can often be encountered around carrion. Nimble flying birds hardly ever fall prey to a sea eagle. But it can happen that ducks or geese with feathers in poor condition (for example, during the molt) are felled by the eagles. Finally, the sea eagle has been known to plunder nests.

High-speed fliers hunt in the open air. Many of the smaller birds of prey are hunting specialists, because of their outstanding skill at flying. The speed records are held by those species with long, pointed wings, such as the falcons. The longer the wings, the greater the distance through which the tip passes at each stroke. The effective airstream is thus very rapid, and the lift developed is considerable.

On the left page: Diving brown pelican (Pelecanus occidentalis). *When swooping down on prey, the bird splashes into the water on its back, braking the dive so that it does not go too deep.*

The Old World kestrel has caught a mouse.

Peregrine falcons scare up other birds from the ground with the express purpose of wearing them down in a long aerial chase and then striking in midair. The falcons called hoobies can even capture swifts and swallows in this way.

More agile birds are not deterred by dense forest or shrubby country in their search for prey. Hawks and kestrels have short, wide wings, the stroke direction of which can be quickly changed, and sturdy tails that help alter the direction of the flight or slow it abruptly. But they are still not maneuverable enough to seize small flying birds; such birds must be caught while they are still. The predator uses its sharp claws like grappling-irons, to hook the victim.

The following observation reveals how this is done. The high warning call of a tit sounded, and the feeding place emptied in the blink of an eye. Fractions of a second before, about 30 small birds had been sitting there, pecking at sunflower seeds. All of them had fled into the bushes. I was very much reminded of tropical fish flitting

Right page: Birds must fly in a way adapted to the number of obstacles in their habitats. Most birds can fly through these dead trees in Lake Nakuru, but only a few can negotiate a dense forest.

Flight silhouettes of birds of prey. From top to bottom: sea eagle, osprey, buzzard, hawk, peregrine falcon, Old World kestrel, sparrow hawk.

between the protective coral branches or the spines of sea urchins when danger threatens. A grey shadow flew up and whipped in a tight curve twice around a leafless, shrubby beech. The little birds in the bush fluttered about with every appearance of terror. The kestrel turned on its side, struck into the bush with one leg, and caught a bewildered sparrow.

Other birds of prey, which feed on slow-moving animals or carrion, need not be so maneuverable. Flying vultures capture living prey only on exceptional occasions. For these birds to find carrion, their chief source of food, it is important to investigate large expanses of ground with minimal expenditure of energy, by soaring. This is made possible by their low sinking speed. The higher they circle, the more territory they have in view. But in order to discover the carrion from such an altitude, their visual acuity must be very great. From an altitude of 3500 meters vultures are said to be able to discern objects only 20 centimeters in diameter! This degree of acuity considerably exceeds that of lower-flying birds of prey.

Animals less mobile than a bird are, of course, particularly vulnerable to attack by birds of prey. Whereas frogs and salamanders depend upon their camouflage and secluded way of life, reptiles that have been warmed up rely more upon the speed of their reactions. To make sure that they can respond quickly, such reptiles in temperate latitudes emerge from their hiding places only during the hot hours of the day. Most small mammals avoid predation by these birds simply by hiding during the day and coming out in search of food only after

nightfall. But even then the world is no paradise for them, since instead of the diurnal birds of prey the owls are on the prowl. Owls can see quite well even at night, and their hearing is excellent. So as not to alert the prey while they are locating and approaching it (and to minimize sound that would mask the sounds of the prey), owls fly silently; this is accomplished by special adaptations. The feathers are very soft, so that sound and air turbulence are kept to a minimum. The leading edges of the wings are frayed in a curious fashion, and the hook barbules of the feathers have been modified into greatly elongated processes that cover each feather with a fluffy coat, providing an acoustical damping layer. Not all owls hunt in twilight or darkness. Some species–for example, the pygmy owl–also hunt by daylight at certain times of the year. These have fewer of the damping feather structures. Such owls rely chiefly upon visual orientation to their prey, so that the noise they make while flying is of less concern.

All the birds that are prey of other birds have been forced to adapt their own manner of flight to this continual threat. How can one escape a nimble, fast assailant with great endurance? Although they are often highly maneuverable, many small birds need places to hide because they tire so quickly. When such birds must migrate across

Below: a pheasant springs into the air (inset) but is caught by a hawk. The picture on the right page shows the hawk eating its prey.

On the following page: Nocturnal owls have feathers modified for sound damping. Top left, toothed leading edge of wing; middle left, barb of a flight feather; and top right, surface of the feather, with long, soft projections from the hook barbules. The two bottom pictures, for comparison, show feathers of pigeons, which are noisy fliers: left, the narrow barb; right, the relatively bare surface of a flight feather.

open country offering no cover, they either travel at night or assemble in flocks.

A cloud of black dots—a flock of starlings—drifts across the scene. It floats to earth, melts into the field, and suddenly separates from the dark background once again. Like a wave it flows over the hedgerows. A large, dark bird appears, a predator. The flock streams toward the new arrival, and the tip of the swarm rounds up to a dense mass. The starlings surge toward the predator and envelope it. Surrounded, the large bird rows vigorously with its wings, lets itself fall and disappears into the darkness below. Like a storm the flock approaches us, and with loud screeches the birds pass over our heads and settle in the trees of a small wood.

That is how a flock of small birds operates. But each member of the flock is protected not only by the active attack behavior just described—a flock offers other advantages. Even if the predator does emerge successful, in the course of a day it may well not capture as many birds from this crowd as it could if it often encountered birds a few at a time. Moreover, a predator finds it much harder to concentrate on one target when confused by a flurry of many identical objects.

I became quite familiar with this "confusion effect" in the course of

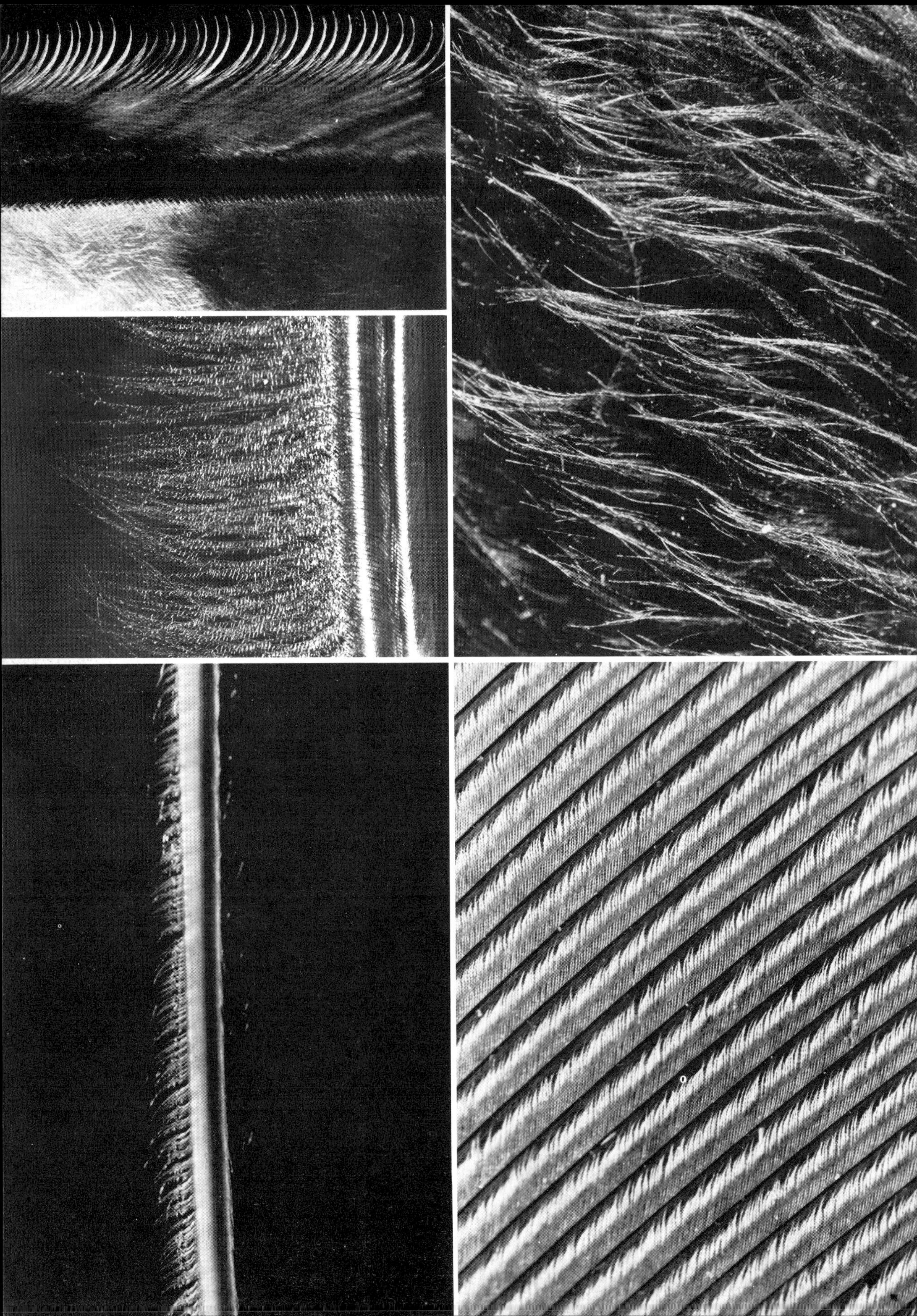

A flock of starlings descending to rest for the night.

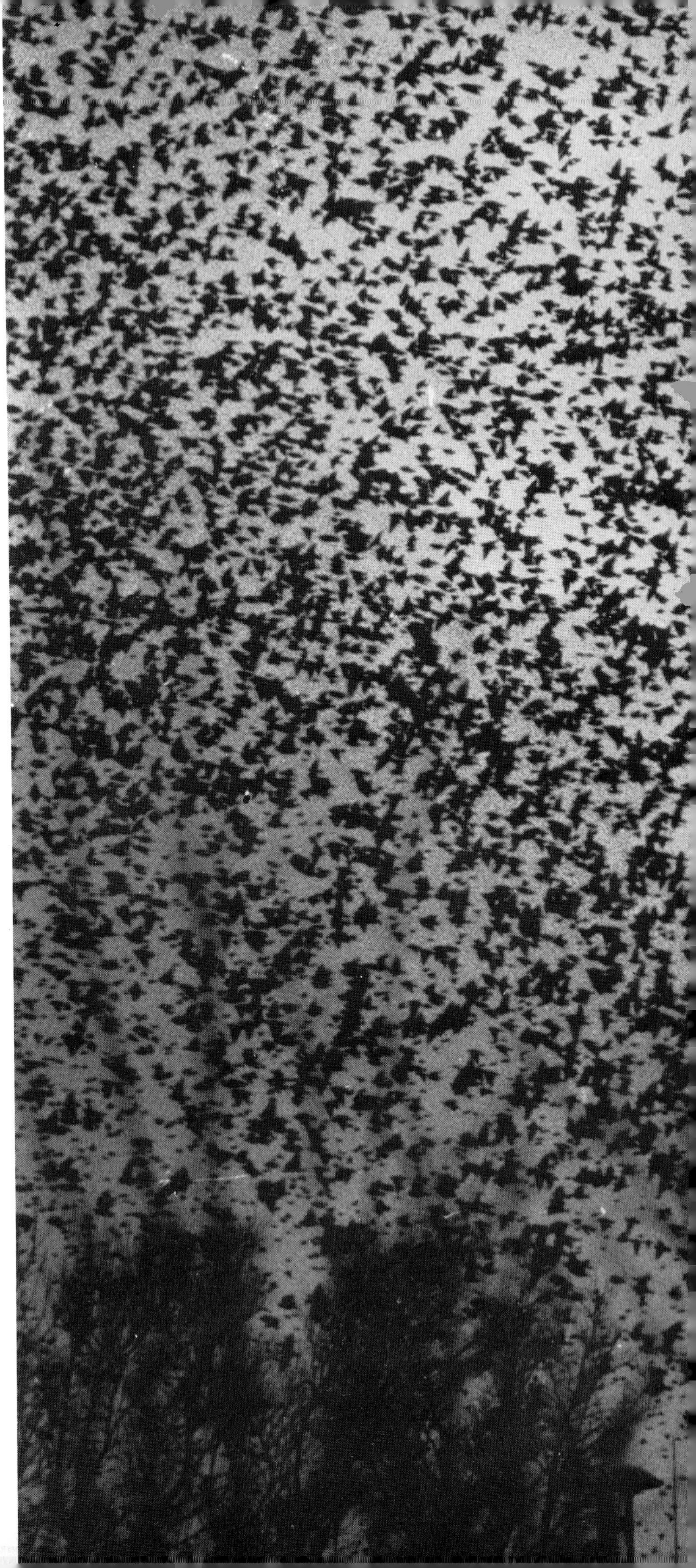

On the preceding page: A barn owl (Tyto alba) *slows down over its prey by increasing the angle of attack; it seizes the prey with its claws and then takes it in its beak, leaving the muscular legs free to catapult it into the air (montage).*

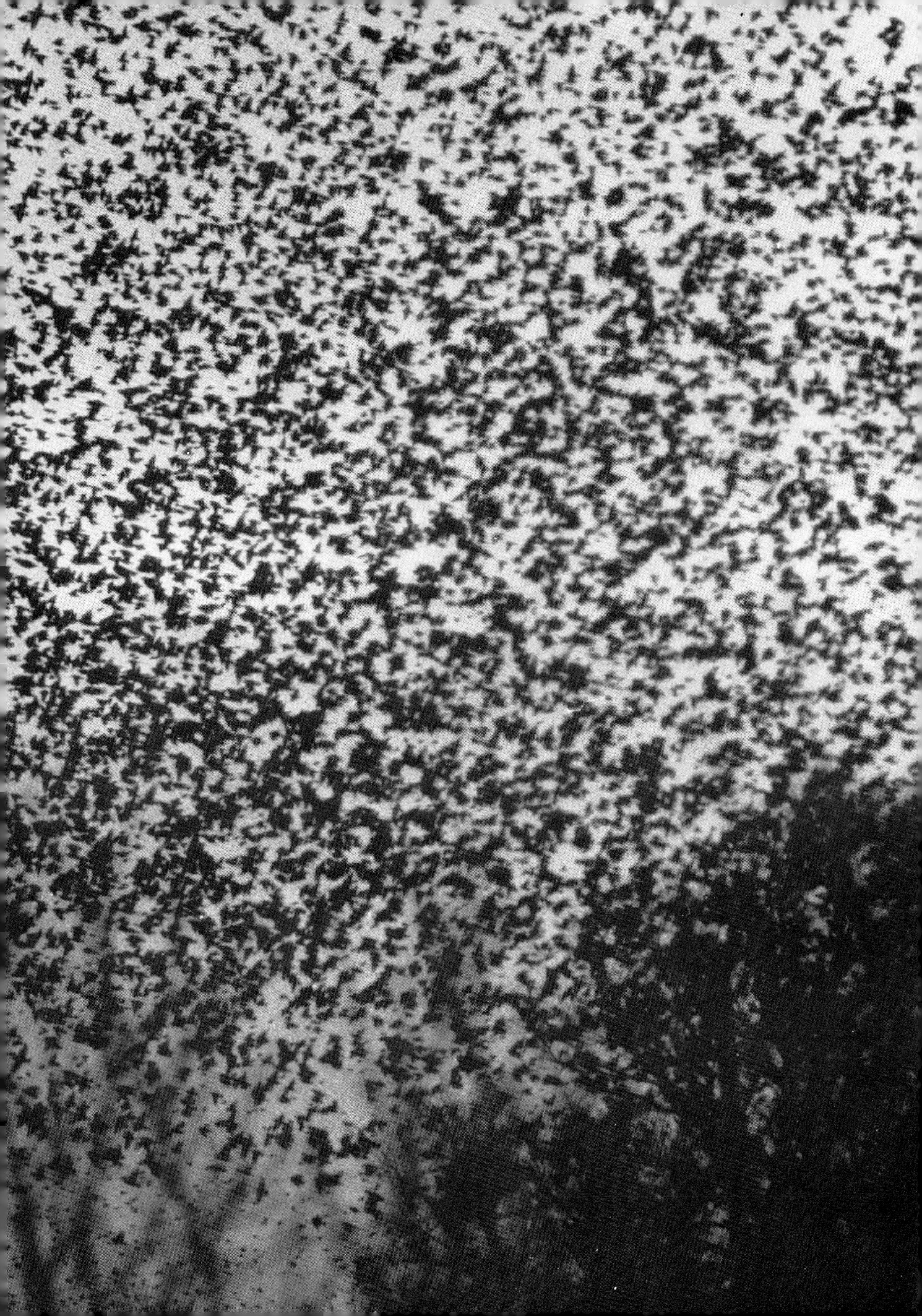

my photographic work. It was always simpler to take pictures of birds flying by themselves than it was to focus on a single member of a group of flying birds. A further advantage for swarming animals is that only those on the fringes can be caught. All the animals in the center of the mass are safe. Small brids are particularly well able to exploit these advantages of flying in a flock. They can follow sudden changes in their neighbors' motion with almost no delay so that the company can maneuver as a unit—the flock becomes a supraindividual entity.

Apart from the flock behavior and agility of small birds, certain birds have evolved a further flight adaptation to protect themselves from avian predators. For a short time after taking off, fowl can fly more rapidly than the birds pursuing them.

When disturbed, pheasants and partridges shoot like bullets toward the nearest trees. It is vital for them to get off to a fast start, since otherwise they could hardly hope to escape. These sprinters are distinguished by the hardness and marked curvature of the primary feathers and the great length of the alula. Their flight musculature is of the type best suited for brief intense work; it is white, without myoglobin. For the first 200 meters the pheasant is carried by the momentum of its explosive take-off, but afterwards a hawk may well

In spite of its long, decorative feathers, a peacock (Pavo cristatus), *like all fowl, is capable of amazingly rapid take-offs (right).*

catch up. If the hunted bird has not found shelter by that time, it is lost. Pheasants can survive only in a habitat with suitable hiding places. In temperate latitudes they have been observed to change habitats regularly. In summer, when the plant cover is dense, a pheasant can slip out of sight almost anywhere and forage for food unobserved. But in winter the leaves fall and the ground cover dies back, so that the pheasant is much more conspicuous. Then it must move to a habitat where it is shielded by forest, thickets, and impenetrable shrubbery. Many pheasant populations leave the open terrain they occupy during the summer and move into safer winter quarters.

The special modifications of flight that adapt birds to one habitat ordinarily prevent their switching to another. But certain birds are equally common in several different biotopes. Laughing gulls can fly over a stormy sea as well as in the still air over the reed masses of inland lakes. Other medium sized flapping fliers such as pigeons and crows can escape from hawks by flying through dense forests and yet remain aloft in a violent tempest over open ground. Such birds are true "generalists."

Although these medium-sized birds as a group have conquered

Left page: Laughing gulls are chasing a coot to rob it of its food. The coot tries to escape by diving (bottom).

more habitats than any other group capable of flight, with a range extending the furthest toward the Poles, they have not succeeded in evolving the largest number of either species or individuals. As regards sheer numbers, the songbirds are the winners by a clear margin. One reason for this success is surely their great maneuverability. Natural forests and brush-covered steppes, where there is an ample food supply, are occupied chiefly by small birds. Since these habitats are full of obstructions to flight, birds that feed there must be agile. Small birds meet this prerequisite and as a result have filled almost all habitats to capacity. Because of their convenient size, they generally can find nesting and hiding places in close proximity to their feeding grounds.

Despite their lack of endurance in flight, some small birds undertake a journey away from their summer habitat in the autumn, when the weather worsens. They have to begin the trip to more favorable regions long before the first frost, so that they can find food on the way. Small birds must stop often to refuel; because their surface areas are large relative to their mass, they rapidly lose both momentum and heat.

Long-Distance Flight

The migration routes of the slender-billed shearwater (Puffinus tenuirostris, *1), the golden plover* (Pluvialis dominica,) *2), the Arctic tern (3), and the white stork (4).*

Arctic terns fly from Spitzbergen to South Africa, golden plovers go nonstop from Alaska to Hawaii, and certain snipe travel from Alaska to Tasmania, evidently crossing 5000 kilometers of ocean without stopping. In winter, many birds in the northern hemisphere are no longer able to find enough food. Grass-eating geese, shore birds which feed on aquatic animals, caterpillar-eaters such as cuckoos, and insect-hunters like flycatchers then move to regions where there is still adequate food—even though the trip is fraught with danger.

After one heavy snowstorm in the midwestern USA, more than 750,000 dead Lapland buntings were found; they had been making their migration. Two hundred dead storks were once washed up on the shores of the Bay of Suez on the Red Sea. In desert regions, exhausted or dead birds are to be found everywhere during the season of migration. Birds of prey have adapted to this seasonal gift of food *en masse.* For example, Eleonora's falcon, which breeds on the rocky islands of the Mediterranean, lays its eggs later than any other bird in this habitat. Egg-laying is put off until it is time for the autumn migration of other birds, for then the situation is so favorable that the falcon can feed its young with no trouble at all. Similarly, the lanner falcon of the Sahara breeds in the spring, when the birds are traveling north again. The falcons of both species have discovered that when

The golden weaver (Textor subaureus) *builds its nest just below that of the sea eagle* Haliaeetus vocifer. *The huge bird of prey cannot harm the small, nimbly flying songbirds; it feeds on fish.*

these small birds, tired by their journey, are flying over terrain that affords no shelter, they are easy prey.

Particularly when their migration takes them across ecologically unfavorable regions such as the Sahara, birds build up a reserve of fat. It is astonishing how quickly they can do this: measurements of a bunting showed that on the first day of fat deposition it increased its body weight by 6.5% Measurements and computations based on various species of small birds indicated an energy consumption of 0.06 to 0.16 kilocalories per gram body weight per hour. This means that a small bird weighing about 25 grams must burn about 0.25 grams of fat per hour of flight. If it covers 50 kilometers during this hour, it will have used up about 0.005 grams of fat per kilometer. Ten grams of reserve fat are enough for a flight over about 2000 kilometers—an enormous efficiency! A bird planning to go a long distance, then, must be fat when it starts out. Whereas the fat content of these small birds outside the migration season is only about 3–10% of the total body weight, in birds preparing for short or medium-length migrations it increases to 40–50%. The fat is stored throughout the body, with the exception of the heart muscle. In the large pectoral muscle it is burned directly, and from the other storage sites it is transported to the muscles in the form of fatty acids.

Migratory behavior may take many forms. In accordance with their way of life and their limited flight skills, for example, the heavily wing-loaded murres make no attempt to fly from the breeding grounds to their winter quarters. They swim to a more favorable wintering spot, through more than 1000 kilometers of water.

Other birds fly long distances in formation; cranes, geese, and pelicans form a flying wedge. They conserve energy by taking advantage of the upward component of the vortices produced by induced drag at the wingtips of neighboring birds, using this as a sort of updraft. But only large birds can be seen flying in such an array. Small birds must beat their wings too rapidly. The quick succession of strokes causes the direction of the air vortices to change continually, so that they are no longer useful.

Another achievement, just as remarkable as their extraordinary metabolic performance, is the orientation ability of these birds. How do they find their goal on a journey of thousands of kilometers, even in the worst weather?

As many as 40 years ago my father did a series of experiments, the results of which are still not explicable today. Feeding starlings were captured at various breeding sites scattered throughout central Europe and shipped to Berlin. There they were all released. Of the 353 birds that started out, 120 returned to their breeding grounds—even though most of them had never before been in the place where they were released. The birds could tell which direction to take from Berlin to find their homes, as though they were using some inborn map.

Once the direction in which they must fly is known, it appears that birds have little difficulty in following it. There are several possible means by which they keep to their course. For example, they could orient by the position of the sun, checked against the internal clock (an innate ability to tell the time). To keep a constant direction, they compensate for the changing position of the sun by continually changing their angle to it. Birds often fly high above the clouds in order to orient by the sun. But how do migrating birds orient at night, or on days when a very high, dense overcast prevents them from seeing the sun at all?

Recent findings indicate that birds can also use the magnetic field of the earth as an aid to orientation. Whether the stars, too, serve as navigational aids is a matter of continuing debate.

Young birds are born knowing the direction in which they should migrate. This ability has been demonstrated by experiments with young hooded crows. Before they began their migration, my father took them a few hundred kilometers away from their natural starting point and set them free there. They flew off in the direction instinct demanded—on a line parallel to that they would otherwise have followed—to new winter quarters.

Learning, however, can cause the birds to modify considerably the innate direction of migration. This is exemplified by the storks, which follow two different routes on their migration to Africa. All the storks living west of the River Elbe go by way of Gibraltar, whereas all those living to the east go *via* the Bosporus. If young storks are taken from populations east of the Elbe and released in the western region in the autumn, after the local storks have already migrated away, the young birds in the new region retain their old, innate flight direction and fly in the direction of the Bosporus. But if the young birds are released while the older birds in the western region are still there, the juveniles join these and migrate with them *via* Gibraltar.

Birds starting out on the long trip for the first time must also have a way of knowing when they have reached their goal—when they should stop flying. To this end, they appear to develop an internally timed "migratory restlessness," which dies out after a certain time has elapsed. Under normal circumstances, such a time will have elapsed when the birds arrive at the place they are seeking, their winter quarters of breeding grounds.

Pelicans flying in formation.

What Distinguishes Birds from People

Snow geese can travel 2700 kilometers in 60 hours. Humans, before they had engines, could manage to go only a tenth as far in the same time. Moreover, men were not as adept as birds at fleeing from the adverse conditions they encountered; they had to find ways of coping with the effects of severe cold, drought, or incessant rain. For independence of a particular source of food that might vanish in time of need, man evolved as an omnivorous animal. Food is always available in some form or another—as roots, fruit, fish, or meat. It was simply a question of being able to take advantage of all these forms. Man developed omnivore dentition and an omnivore digestive system. His hands and the tools he invented helped him capture and prepare food. The development of methods of storing food, continually refined hunting techniques, and the need for protection against predaceous animals required a steady evolution of new behavioral patterns. Men formed groups, sought out caves, and devised weapons. In so doing they put themselves in a position of dependence upon their own inventions but became more independent of their environment.

Birds, which can quickly leave unfavorable habitats, have not been compelled to interact with their environment in such a way that intelligent actions have been necessary for survival. By contrast, man owes his evolution to this very need to apply his intelligence. His continued survival depends upon the way he applies it, now and in the future.

Glossary

Aerodynamics. The science of the movement of gases, in particular, the air. A subdiscipline within fluid dynamics.

Airstream. The current of air flowing over and near the wing surface. The net airstream at any instant is determined both by the motion of the body with respect to the air and by motions of the wings with respect to the body. Rapid changes in the wind also alter the airstream. During a wingstroke the direction of the airstream changes continually. Behind the wing the air no longer flows in the direction it does over the wing surface but is deflected downward (in reaction to the net upward impulse given to the bird).

Angle of attack. The angle between the plane of a wing (more precisely, the chord, q.v.) and the direction of the airstream. It is also called the aerodynamic angle of incidence. The "critical angle of attack" is that at which maximal lift is generated and the airstream just begins to separate from the wing surface.

Aspect ratio. In aeronautical engineering this is the ratio of wing-span—i.e., the distance between the two wingtips—to chord length (q.v.). The term is also applied to bird wings in this book, with the difference that here it refers to the base-to-tip length and average chord of a single wing.

The advantage of a large aspect ratio is that induced drag is small and lift relatively large, and the bird is not easily upset by gusts of wind. The disadvantage is that very long wings cannot be moved as rapidly as short wings, since particularly long oscillating wings have a large mass that is subject to the laws of inertia. As they beat, the wings must repeatedly be slowed down and accelerated again, which costs time and energy.

Bernoulli's Law. Throughout a flowing fluid at approximately constant altitude (i.e., constant potential energy), static pressure (P) plus dynamic pressure (the kinetic energy per unit volume) is constant. Over the convex upper surface of a wing, velocity (and hence dynamic pressure) is increased so that static pressure is reduced, resulting in net lift.

Boundary layer. The particles of air immediately adjacent to a body tend to adhere to its surface and thus to slow down air particles flowing immediately above them. The layer of the airstream in which this braking action occurs is the boundary layer. The air in this layer can flow in a laminar or turbulent manner. A laminar boundary layer generates less drag but can more readily induce separation of the main airstream from the wing surface at low airstream velocity and high angles of attack; in such conditions a turbulent boundary layer can still adhere, preventing separation. Above Reynolds numbers (q.v.) in the region of 10^5, a boundary layer tends to become turbulent (in measurements using artificial airfoils).

Center of pressure. The center of pressure is the point at which the resultant of the aerodynamic forces acting on a wing intersects the chord line of the wing. If the center of pressure and the center of gravity of a bird are aligned, one prerequisite for stable flight is fulfilled. If the center of pressure is behind the center of gravity, there is a tendency for the bird to pitch head down; in the reverse situation, the bird can pitch head up.

Chord. As used here, the length of the straight line connecting leading and trailing edges of a wing, at any given distance from the wing base.

Drag. The net force resisting forward motion. It may be considered as having components either independent of lift (profile drag) or varying with lift (induced drag). From another point of view it is considered as having friction (also called form drag) and pressure (also called viscous drag) components. The total drag is smaller, the smaller the area presented to the relative wind, the smoother the surface, and the more sharply pointed (given a large aspect ratio) the wings. An increase in pressure drag is associated with reduction of lift and in some cases with separation of the airstream from the wing.

Glide ratio. The ratio of forward speed to sinking speed (or the corresponding distances per unit time) during a glide. The larger the glide ratio, the smaller the glide angle—that is, the more lift and the less drag generated. The ratio of lift to drag is also used to describe flying objects.

Inertia. The tendency of every mass to resist acceleration and deceleration. The implication in bird evolution is that heavy birds and wings require greater forces to slow down or accelerate than lighter ones.

Kinetic energy. The energy (capacity to do work) derivable from a moving mass by virtue of the fact that it is moving. The kinetic energy per unit volume of flowing air is given by $\frac{1}{2}Þv^2$, where Þ is density and v its velocity. The dimensions are the same as those of pressure, as Bernoulli's Law implies; thus $\frac{1}{2}Þv^2$ is often called dynamic pressure.

Landing flaps. Movable surfaces (either parts of the trailing edge of the wing which can be bent down or separate retractable sections which can be extended when required) used to increase the effective curvature of the wing during slow flight. This increase counteracts, to a certain extent, the effect of the lower airstream velocity, so that lift remains adequate.

Lift. In most aerodynamic usage, lift is the force generated (by static pressure differential) perpendicular to the relative wind incident upon a wing. Its magnitude changes as the square of the airstream velocity and also depends upon the angle of attack, the presence and type (laminar or turbulent) of a boundary layer, the wing chord, the wing length, and the shape of the wing (or of the feathers, each of which can also generate lift if it is suitably arched). The net upward force on a flying object results from a combination of lift, drag, and gravitational forces. In bird flight the relationships are especially complex (see text).

Polar diagram. A means of summarizing the forces acting on a wing. The coefficient of lift C_L is represented by the ordinate and the coefficient of drag C_D by the abscissa, for each of many different angles of attack.

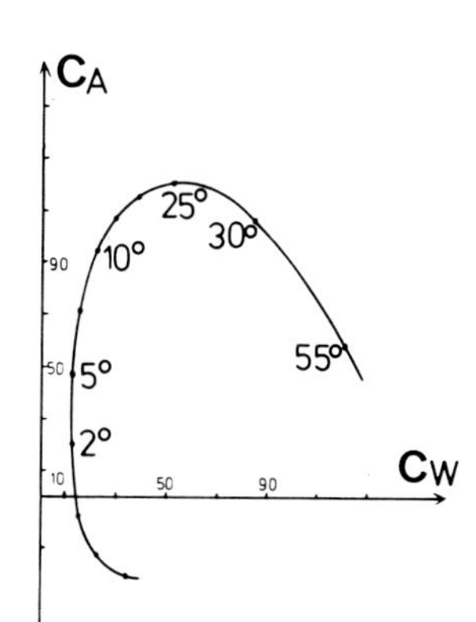

The diagram permits the comparison of profiles of different size, since the area of the wing is incorporated in the formulae for the coefficients. The polar diagram tells which angle of attack is associated with the best combination of required lift and tolerable drag; the compromise affects the gliding angle and the position of the center of pressure.

Profile. The shape of the cross section of a wing. Its general properties are a more or less rounded leading edge, a convex-upward curvature, and a sharp trailing edge. As a rule, bird-wing profiles (especially in the parts of the wing distant from the body) have a small radius of curvature at the leading edge, small thickness, and variable curvature.

Reynolds number. An indicator of the relative magnitudes of inertial and viscous forces in fluid currents. High Reynolds numbers correspond to a predominance of inertial influences, while with low Reynolds numbers viscosity predominates. In general, boundary layers change from laminar to turbulent as Re increases above 10^5. Since a turbulent boundary layer is important to flying birds, which often use high angles of attack, they must either fly in the range of high Re or make the boundary layer turbulent in other ways (cf. turbulence generators).

Separation point. A point along the wing profile at which the boundary-layer air over a wing has slowed sufficiently by internal friction to allow reverse flow from the trailing edge and the development of vortices behind the separation point; the main airstream is then separated from the wing surface by this region of "burbling." The presence of a turbulent (rather than laminar) boundary layer can delay the onset of separation and narrow the turbulent wake.

Supplementary lift generators. These are accessory devices to ensure that even when wings are at a large angle of attack and the airstream is slow, lift continues to be produced. Examples are landing flaps, subsidiary airfoils in front of the leading edge, and slotted wings.

Turbulence generators. Sharp edges, structures projecting from the wing surface, and surface roughness can render a boundary layer turbulent even at low Reynolds numbers. There are probably a great many structural specializations of bird wings and feathers that serve this purpose.

Velocity of flight. A decisive factor in lift and drag production is the velocity of a bird with respect to the air, while from the point of view of locomotion the ground speed is important. The bird can alter its speed in several ways: in gliding, by increasing wing loading, and in horizontal flight by raising the wingbeat frequency. Birds with long pointed wings are particularly well able to reach high flight speeds since the outer parts of the wings, beating through a wide arc, move rapidly with respect to the air and generate large thrust. Some examples of flight speed in km/hr (rough estimates only):

Bird	km/hr	Bird	km/hr
Peregrine falcon (swooping on prey)	300	Starling	80
		Cormorant	70
Peregrine falcon (horizontal flight)	80	Barn swallow	60
		Sparrow	50

Swift, horizontal	90	Stork	50
Pigeon (maximum)	90	Magpie	45
Northern gannet	80		
Imperial sandgrouse	80		

(from V. Meinertzhagen and others)

Warping. Twisting of a wing about its long axis, as exhibited in bird wings during the downstroke. The angles of attack are made smaller toward the tip of the wing. Bird wings are also permanently warped, in the sense that the profile changes along their length. The consequence of warping is that the aerodynamic forces in the outer part of the wing are directed more forward (contributing to thrust) than in the proximal part, the latter having a higher angle of attack.

Wing characteristics. Critical characteristics of a wing include its geometry (size, profile, warping, outline) and the physical properties of the material of which it is made (flexibility, elasticity, stiffness, deformability, and weight of its various parts). These characteristics differ widely among bird species and are adapted to the manner of flight.

Wing loading. The ratio between the weight of a bird and the area of the aerodynamic bearing surfaces. Wing loading is sometimes expressed with reference to the wing area only, but it is useful to include the effective areas of body and tail. The range of variation can be appreciable; variable factors include the nutritional state of the bird (differences in weight of 30% are no rarity; in fasting experiments we have observed weight losses of up to 50%). There are many possible combinations of flexion or extension of wings and tail by means of which a bird can change its wing loading (aircraft can do so only if equipped with retractable lift-generating surfaces).

Some examples of wing loading (in g/dm^2):

1. Thick-billed murre	260
2. Albatross	168
3. Northern gannet	110
4. Mallard	112
5. Stork	67
6. Homing pigeon	57
7. Ordinary domestic pigeon	45

8. Buzzard	42
9. Blackbird	30
12. Bullfinch	22
11. Siskin	19
12. Great tit	17

These area measurements include both wings and tail. The data for Numbers 4, 6, 7, 9, 10, 11 and 12 are mean values for several animals, measured at different times of day.

Acknowledgments

This book could not have been produced without the collaboration of my wife, Gerda-Maria. An inexhaustible companion, unfailing source of encouragement, and tireless proofreader, she worked toward the realization of our shared idea—and often accepted uncomplainingly, considerable financial sacrifices. My thanks are also due to Hans Scherz, who helped to adapt the text and the basic plan of the book.

Finally, Prof. G. Wolf, of the Institute for the Scientific Film in Göttingen, and Prof. H. Remmert, of the Second Zoological Institute, Erlangen, assisted our undertaking in many ways.

The generous format of the book, with its large-scale reproductions of the photographs, was made possible in the original German edition by the publishers, Kindler Verlag. That edition profited greatly from the critical reading of the manuscript by Dr. H. Oehme and from discussions and improvements of my simplifications of the physics by Dr. D. Ronneberger.

Sources of the Illustrations

Most of the photos reproduced here were taken with a Leicaflex mot camera with Telyt 5.6 mm lens, using Ilford FP4 or Ektachrome High Speed film. The shortest possible exposure time was always used (usually 1/1000 or 1/2000 sec). I took the pictures of owls in a flight cage, with my friend Dr. D. Haarhaus, using a two-lens mirror reflex camera and the flash apparatus Computerblitz Mecablitz 402. I used the same apparatus for the photographs of Old World kestrels and chaffinches, which were made in the laboratory. Whereas two flash lamps were used for all these electronic flash photos, the pictures of great tits taken outdoors employed only a single small Computerblitz device. The condor, the bearded vulture, the greylag geese, and the peacock were photographed in the Nuremberg Zoo.

The stereoscan photos of feather surfaces were provided by Dr. Klingele and the airplane photos, by Lufthansa: The films of the redstart taken for slow-motion projection were made outdoors with help from the Institute for the Scientific Film in Göttingen, which also provided assistance with the stroboscope pictures of the parakeet. All the other photographic illustrations were made on my trips to Spitzbergen, Africa, and the Galapagos, with a Bolex 16 Reflex.

Other Illustrations

P. 12/13 from Schmidt, *Flug der Tiere*; p. 19 from Burian; p. 20 from "Homo ludens" Nr. 20, "Simplizissimus," Historical Picture Archive Handke, Archive for Art and History; pp. 22–27 from original drawings by the authors indicated; pp. 30 and 34 from Naumann-Hennicke; p. 31 bottom from Sy and Stolpe; p. 33 sketch from Stolpe; p. 39 bottom from Herzog; p. 51 sketches from Oehme; p. 60 from Dubs; p. 96/97 from E1618 and E1617, Films, Institute for the Scientific Film; p. 126, Ullstein Bilderdienst.

Scientific Films

The following films can be obtained on loan from the Institut fur den Wissenschaftlichen Film, Nonnenstieg 72, Göttingen, Germany.

INST. F. D. WISS. FILM: E 1618: *Chlorostilbon melanorhynchus* (Kolibri)–Richtungsänderungen beim Flug, Göttingen, 1971. (Directional changes in hummingbird flight.)

–E 1617: *Chlorostilbon melanorhynchus* (Kolibri)–Flug auf der Stelle, Göttingen, 1971. (Hovering by the hummingbird.)

RÜPPELL, G.: E 1507: *Uria lomvia* (Dickschnabellumme)–Landung am Brutfelsen, Göttingen, 1969. (Thick-billed murres landing on the breeding rock.)

–E 1619: *Fulmarus glacialis* (Eissturmvogel)–Landung am Brutfelsen, Göttingen, 1970. (Fulmars landing on the breeding rock.)

–E 1844: *Phoenicurus phoenicurus* (Gartenrotschwanz)–Flugmanöver, Göttingen, 1972. (Flight maneuvers by redstarts.)

–E 2181: *Pelecanus occidentalis* (Meerespelikan)–Stosstauchen, Göttingen, 1974. (Brown pelicans diving.)

–E 2182: *Sula nebouxii* (Blaufusstölpel)–Stosstauchen, Göttingen, 1974. (Blue-footed boobies diving.)

STOLPE, M. and K. ZIMMER.: E 632: *Cinnyris senegalensis* (Nektarvogel)–Trillerflug, Göttingen, 1964. ("Trill flight" by sunbirds.)

References

ALEXANDER, R. McN.: Muscle performance in locomotion and other strenuous activities. In L. Bolis, S.H.P. Maddrell, and K. Schmidt-Nielsen, (eds.): Comparative Physiology, pp. 1-21. North-Holland Publ., Amsterdam. 1973.

ALEXANDER, W.G.: *Die Vögel der Meere.* Parey Verlag, Hamburg. 1959.

AYMAR, G.: *Herrlicher Vogelflug.* Kresler u. Co. 1949.

BERGER, M.: Energiewechsel von Kolibris beim Schwirrflug unter Höhenbedingungen. *J. Ornithol.* **115**:273-288, 1974.

BERGER, M. and J.S. HART.: Physiology and energetics of flight. In D.S. Farner and J.R. King: *Avian Biology IV,* pp. 415-477. Academic Press, New York. 1974.

BERNSTEIN, M.H., S.P. THOMAS and SCHMIDT-NIELSEN: Power input during flight of the Fish Crow *Corvus ossifragus. J. Exp. Biol.* **58**:401-410, 1973.

BILO, D.: Flugbiophysik von Kleinvögeln. I. Kinematik und Aerodynamik des Flügelabschlags beim Haussperling (Passer domesticus L.). *J. Comp. Physiol.* **71**:382-454, 1971. II. Kinematik und Aerodynamik des Flügelaufschlags beim Haussperling (Passer domesticus L.). *J. Comp. Physiol.* **76**:426-437, 1972.

BÖKER, H.: Die biol. Anatomie der Flugarten der Vögel und ihre Phylogenie. *J. Ornithol.* **75**:304-371, 1927.

BÖKER, H.: Einführung in die vergleichende biologische Anatomie der Wirbeltiere. Band I, II. Fischer, Jena. 1935 and 1937.

BRAMWELL, C.D. and G.R. WHITFIELD: Biomechanics of *Pteranodon. Phil. Trans. Roy. Soc. Lond. B* **267**:503-581, 1974.

BROWN, R.H.J.: The flight of birds. II. Wing function in relation to flight speed. *J. Exp. Biol.* **30**:90-103, 1953.

BROWN, R.H.J.: The power requirements of birds in flight. *Symp. Zool. Soc. Lond.* **5**:95-99, 1961.

BROWN, R.H.J.: The flight of birds. *Biol. Rev.* **38**:460-489, 1963.

CONE, C.D.: Thermal soaring in birds. *Am. Scientist* **50**:180-209, 1962.

CONE, C.D.: A mathematical analysis of the dynamic soaring flight of the albatross with ecological interpretations. *Va. Inst. Mar. Sci., Spec. Sci. Rep.* **50**:1-104, 1964.

DEMOLL, R.: Die Flugbewegungen bei grossen und kleinenVögeln. *Z. Biol.* **90**:199-230, 1930.

DUBS, F.: *Aerodynamik der reinen Unterschallströmung.* Birkhäuser Verlag, Basel. 1966.

DUNCKER, H.-R.: Das Lungen-Luftsacksystem der Vögel. Ein Beitrag zur funktionellen Anatomie des Respirationsorgans. Habilitation Paper, Med. Fak., Univ. Hamburg. 1968.

ENGELS, F.M.: *So bewegen sich die Tiere auf dem Land, im Wasser, in der Luft.* Südwest Verlag, Munich. 1969.

GLUTZ v. BLOTZHEIM, U.: *Handbuch der Vögel Mitteleuropas.* 5 Vols. 1969–1973.

GRAY, J.: *Animal Locomotion.* Weidenfeld & Nicolson, London. 1968.

GREENEWALT, C.H.: *Hummingbirds.* Doubleday, New York. 1960.

GREENEWALT, C.H.: Dimensional relationships for flying animals. Smithson. Misc. Collect. 144, No. 2, 1962.

GROEBBELS, F.: Der Bauplan des Vogels und das Flugproblem. *Pflüg Arch.* **212**:215-228, 1926.

GRZIMEK, B.: *Grzimek's Animal Life Encyclopedia.* Van Nostrand Reinhold, New York. 1974–1976.

HAUBENHOFER, M.: Die Mechanik des Kurvenfluges. *Schweiz. Aerorev.* **39**:561-565, 1964.

HECK, H.D.: *Der Flug.* Suhrkamp Verlag, Frankfurt/Main. 1971.

HEINROTH, O. and M.: *Die Vögel Mitteleuropas.* Berlin. 1969.

HERTEL, H.: *Biologie und Technik. Struktur, Form, Bewegung.* Krausskopf, Mainz. 1963.

HERZFELD, M.: *Leonardo da Vinci als Denker, Forscher und Poet. Aus seinen veröffentlichten Schriften.* Diederichs, Jena. 1926.

HERZOG, K.: Anatomie und Flugbiologie der *Vögel.* Fischer, Stuttgart. 1968.

HILL, A.V.: The dimensions of animals and their muscular dynamics. *Sci. Progr. (London)* **38**:209-230, 1950.

v. HOLST, E. and D. KÜCHEMANN: Biologische und aerodynamische Probleme des Tierflugs. *Naturwiss.* **29**:348-362, 1941.

v. HOLST, E.: Über "Künstliche Vögel" als Mittel zum Studium des Vogelflugs. *J. Ornithol.* **91**:406-447, 1943.

HUMMEL, D.: Die Leistungsersparnis beim Verbandsflug. *J. Ornithol.* **114**:259-282, 1973.

IDRAC, P.: *Experimentelle Untersuchungen über den Segelflug.* Oldenbourg Verlag, Munich. 1932.

KIPP, F.A.: Der Handflügel-Index als flugbiol. Mass. *Ibis* **20**:77-86, 1960.

LILIENTHAL, O.: *Der Vogelflug als Grundlage der Fliegekunst.* Gaertners Verlagsbuchhandlung, Berlin. 1889.

LORENZ, K.: Beobachtetes über das Fliegen der Vögel und über die Beziehungen der Flügel- und Steuerform zur Art des Fluges. *J. Ornithol* **81**:107-236, 1933.

LORENZ, K.: *Die "Erfindung" von Flugmaschinen in der Evolution der Wirbeltiere, in: "Darwin hat recht gesehen".* Neske Verlag. 1965.

MAREY, E.J.: *Physiologie du mouvement. Le vol des oiseaux.* Masson, Paris. 1890.

MCGAHAN, J.: Gliding flight of the Andean condor in nature. *J. Exp. Biol.* **58**:225-237, 1973.

MCGAHAN, J.: Flapping flight of the Andean condor in nature. *J. Exp. Biol.* **58**:239-253, 1973.

MCHUGH, T.: Wings under Water. *Nat. Hist. New York* **61** and **69**, pp. 160-163, 1952.

MEINERTZHAGEN, R.: The speed and altitude of bird flight. *Ibis* **97**:81-117, 1955.

MEISE, W. and R. BERNDT.: *Naturgeschichte der Vögel.* 3 Vols. Franckh'sche Verlagshandlung, Stuttgart. 1966.

MUYBRIDGE, R.: *Animals in Motion.* London. 1899. New edition by Dover Publications, New York. 1957.

NACHTIGALL, W. and J. WIESER.: Profilmessungen am Taubenflügel. *J. Comp. Physiol.* **52**:333-346, 1966.

NACHTIGALL, W.: *Insects in Flight.* George Allen & Unwin, London. 1974.

Nachtigall, W. and B. Kempf.: Vergleichende Untersuchungen zur flugbiologischen Funktion der Alula spuria ("Daumenfittich") bei Vögeln. I. Der Daumenfittich als Hochauftriebserzeuger. *J. Comp. Physiol.* **71**:326-341, 1971.

Nachtigall, W.: Geschichte der Erforschung des Vogelfluges von der Renaissance bus zur Gegenwart. *J. Ornithol.* **114**:283-304, 1973.

Nachtigall, W.: *Phantasie der Schöpfung.* Hoffmann u. Campe, Hamburg. 1974.

Nachtigall, W.: *Biophysik des Tierflugs: Rheinisch-Westfälische Akademie der Wissenschaften.* Westdeutscher Verlag, Düsseldorf. 1974.

Nicolai, J.: *Vogelleben.* Belser Verlag, Stuttgart. 1973.

Odum, E.P., C.E. Connel and H.L. Stoddard.: Flight energy and estimated flight ranges of some migratory birds. *Auk* **78**:515-527, 1961.

Oehme, H.: Untersuchungen über den Flug und Flügelbau von Kleinvögeln. *J. Ornithol.* **100**:363-396, 1959.

Oehme, H.: Der Kraftflug der Vögel. *Vogelwelt* **89**:20-41, 1968.

Oehme, H.: Über besondere Flugmanöver des Mauerseglers. *Beitr. zur Vogelkunde* **13**(6):393-396, 1968.

Oehme, H.: Der Rüttelflug des Gartenrotschwanzes. *Beitr. zur Vogelkunde* **15**(6):417-433, 1970.

Oehme, H.: Über die geometrische Verwindung des Vogelflügels. *Biol. Zentralblatt* **90**:145-156, 1971.

Parrott, G.C.: Aerodynamics of gliding flight of a Black Vulture *Coragyps atratus. J. Exp. Biol.* **53**:363-374, 1970.

Pennycuick, C.J.: Gliding flight of the Fulmar Petrel. *J. Exp. Biol.* **37**:330-338, 1960.

Pennycuick, C.J.: Gliding flight of the White-backed Vulture *Gyps africanus. J. Exp. Biol.* **55**:13-38, 1971.

Pennycuick, C.J.: Control of gliding angle in Rüppell's Griffon Vulture *Gyps rüppellii. J. Exp. Biol.* **55**:39-46, 1971.

Pennycuick, C.J.: Gliding flight of the dog-faced bat *Rousettus aegyptiacus* observed in a wind tunnel. *J. Exp. Biol.* **55**:833-845, 1971.

Pennycuick, C.J.: *Animal Flight.* Studies in Biology No. 33. Arnold, London. 1972.

Pennycuick, C.J.: The soaring flight of vultures. *Sci. Am.* **229**(6): 102-109, 1973.

Pennycuick, C.H.: Mechanics of flight. In D.S. Farner and J.R. King: *Avian Biology V.,* pp. 1-75. Academic Press, New York. 1975.

Penzlin, H.: *Kurzes Lehrbuch der Tierphysiologie.* VEB Gustav Fischer Verlag, Stuttgart-Hohenheim. 1970.

Peterson, R., G. Mountfort and P.A.D. Hollom.: *Die Vogel Euröpas.* Parey Verlag, Hamburg. 1973.

Raspet, A.: Biophysics of bird flight. *Science* **132**:191-200, 1960.

Rüppell, G.: Flugmanöver des Gartenrotschwanzes (Phoenicurus phoenicurus L.). *J. Comp. Physiol.* **71**:190-200, 1971.

Rüppell, G.: Flugbiologische Anpassungen felsbrütender Vögel auf Spitzbergen. *Natur und Museum* **101**:69-76, 1971.

Rüppell, G.: Aerodynamisch bedeutsame Strukturen am bewegten Kleinvogelflügel. *J. Ornithol.* **114**:220-226, 1973.

Schack, W., O. Leege and H. Focke.: *Wunder des Möwenfluges.* Bechthold Verlag. 1937.

SCHEITHAUER, W.: *Kolibris–fliegende Edelsteine.* Bayer. Landwirtschafts-Verlag.
SCHMIDT, H.: *Der Flug der Tiere.* Verlag Waldemar Kramer. 1960.
SCHMIDT-NIELSEN, K.: Locomotion: energy cost of swimming, flying, and running. *Science* **177**:222-228, 1972.
SCHÜZ, E.: *Grundriss der Vogelzugskunde.* Parey Verlag, Hamburg. 1971.
SEELMANN, D.: *Fliegen–Gestern–heute–morgen.* Buch und Zeit Verlag. 1970.
SLIJPER, E.J.: *De vliegkunst en het dierenrijk.* Brill, Leiden. 1950.
SLIJPER, E.J.: *Riesen und Zwerge im Tierreich.* Parey Verlag, Hamburg. 1960.
STOLPE, M. and K. ZIMMER.: Der Schwirrflug des Kolibris im Zeitlupenfilm. *J. Ornithol.* **87**:136-155, 1939.
STOLPE, M. and K. ZIMMER.: *Der Vogelflug.* Akad. Verlagsgesellschaft, Leipzig. 1939.
STORER, J.H.: Bird aerodynamics. *Sci. Am.* **186**:24-29, 1952.
STORER, J.H.: Weight, wing area and skeletal proportions in three Accipiters. *Acta XI Congr. International Ornithol. Basel,* pp. 287-290, 1954.
STRESEMANN, E. and VESTA.: Die Mauser der Vögel. *J. Ornithol.* **107**: , 1966.
SY, M.: Funktionell-anatomische Untersuchungen am Vogelflügel. *J. Ornithol.* **84**:200-296, 1936.
TUCKER, V.A.: Oxygen consumption of a flying bird. *Science* **154**:150-151, 1966.
TUCKER, V.A.: Flight energetics in birds. *Amer. Zool.* **11**:115-124, 1971.
TUCKER, V.A.: Metabolism during flight in the laughing gull, *Larus atricilla. Am. J. Physiol.* **222**:237-245, 1972.
TUCKER, V.A.: Bird metabolism during flight: Evaluation of a theory. *J. Exp. Biol.* **58**:689-709, 1973.
TUCKER, V.A. and G.C. PARROTT.: Aerodynamics of gliding flight in a falcon and other birds. *J. Exp. Biol.* **52**:345-367, 1970.
VAUCHER, C.: *Vögel im Flug.* Fretz & Wasmuth Verlag. 1964.
WEIS-FOGH, T.: Energetics of hovering flight in hummingbirds and in Drosophila. *J. Exp. Biol.* **56**:79-104, 1972.
WEIS-FOGH, T.: Quick estimates of flight fitness in hovering animals, including novel mechanisms for lift production. *J. Exp. Biol.* **59**:169-230, 1973.
WEIS-FOGH, T. and M. JENSEN.: Biology and physics of locust flight. I. Basic principles in insect flight. A critical review. *Phil. Trans. Roy. Soc. Lond. B* **239**:415-458, 1956.
WOOD, C.J.: The flight of albatrosses (a computer simulation). *Ibis* **115**:244-256, 1973.

Index

Asterisk indicates pages on which illustrations appear.